Genetics of Domestications

SCIENCES

Biology, Field Director – Marie-Christine Maurel

Genetics, Epigenetics, Subject Head – Bernard Dujon

Genetics of Domestications

Coordinated by
Georges Pelletier

iSTE

WILEY

First published 2024 in Great Britain and the United States by ISTE Ltd and John Wiley & Sons, Inc.

ISTE Ltd
27-37 St George's Road
London SW19 4EU
UK

www.iste.co.uk

John Wiley & Sons, Inc.
111 River Street
Hoboken, NJ 07030
USA

www.wiley.com

Library of Congress Control Number: 2024935671

British Library Cataloguing-in-Publication Data
A CIP record for this book is available from the British Library
ISBN 978-1-78945-167-2

ERC code:
LS2 Genetics, 'Omics', Bioinformatics and Systems Biology
LS2_5 Epigenetics and gene regulation
LS3 Cellular and Developmental Biology
LS3_9 Developmental genetics in animals and plants

Contents

Introduction

Michèle TIXIER-BOICHARD[1] **and Georges PELLETIER**[2,3]
[1]*GABI, INRAE, AgroParisTech, Université Paris-Saclay, Jouy-en-Josas, France*
[2]*Académie des sciences, Paris, France*
[3]*Académie d'agriculture de France, Paris, France*

Based on a definition

Domestic, in the first sense, means belonging to the house, in Latin "***domus***". Thus, for a living being, it means living in a home, in the broadest sense, of man, being raised and fed there. The domestication of animals, plants and microorganisms is therefore a process of adapting individuals, then their descendants, to an environment that is modified by human groups. This adaptation is associated with hereditary transformations that are conducive to their exploitation, which are expressed, for example, in an animal's docile nature or the limitation of the spontaneous dispersal of cereal grains. In some cases, the species disappears in the wild and is only represented by domesticated types. Domestication can thus be seen as evolutionary pressure. A generic definition of ***domestication*** has been proposed by the Convention on Biological Diversity[1] as follows: "'Domesticated or cultivated species' means species in which the evolutionary process has been influenced by humans to meet their needs".

The transition from the wild to the domesticated state is just a starting point, because as human needs continue to evolve, so does the adaptation and management of domesticated species. Domestication can thus be seen as a process of accumulating heritable modifications to the characteristics of a group of individuals,

1 Available at: https://www.cbd.int/convention/.

Genetics of Domestications,
coordinated by Georges PELLETIER.

enabling them to better meet the needs of our species, not only for survival, but also for pleasure, or for cultural or symbolic reasons. These genetic transformations of the species we have chosen, and the history of their exploitation for a variety of needs, have led to a diversification of animal breeds and plant varieties, in addition to strains of microorganisms involved in the manufacture of many fermented foods, such as beer, wine, bread, cheese and more. The history of their domestication has become increasingly well documented since the advent of modern DNA analysis technologies, revealing a plethora of strains and species whose ability to evolve and exchange genes contributes to the quality of these products.

Understanding domestication

When and where did domestication begin? How did it happen?

The history of domestication is intimately linked to human history or, more accurately, to the history of different human groups scattered across the planet. Archaeological digs enable us to reconstruct the past of certain species by uncovering their remains and using all the technical resources of archaeobotany and archaeozoology,[2] from morphological description to various types of analysis. These include, for example, carbon isotope content for dating, or DNA sequences that can be extracted to reveal their relationship with today's domesticated species.

In terms of the evolution of modern man, whose emergence can be traced back some 300,000 years to Africa, before migrating to the Eurasian continent and later to Australia and the Americas, the first domestications only appeared very late, in the last two tens of thousands of years before us: dog, man's oldest companion, is thought to have been domesticated for at least 25,000 years. Over the last 12,000 years, plant species have undergone new types of genetic selection, corresponding to the needs of populations and leading to the emergence of agricultural societies in various parts of the world (Purruganan and Fuller 2009). Up until then, man had already been using plant and animal resources through gathering and hunting. Thus, the use of selectively harvested wild cereals, which spontaneously seeded and invaded areas occupied by populations, shows a first stage of co-evolution between man and a species used to make food. Evidence of this can be found in the Paleolithic period, like the case of millet in northern China 28,000 years ago, or barley and wheat, with potential signs of cultivation 23,000 years ago by hunter-gatherers who set up sedentary sites in the Middle East. The beginnings of bread-making are confirmed to have started 14,600 years ago, well

2 Analysis and interpretation of plant and animal remains found on archaeological sites.

before the domestication of wheat. The satisfaction of these needs led to new economic activity involving work in the fields, and the storage and processing of harvests. The seasonal nature of cereals led to the construction of storage facilities. The granaries that precede the emergence of domestication by at least 1,000 years represent a critical change in the relationship between humans and plant foods, with new social organizations of sedentary communities and the emergence of agriculture (Willcox and Stordeur 2012; Vigne 2015).

Different species have been involved in different regions on all continents, identified as hotbeds of domestication. Examples include the Fertile Crescent in the Middle East (wheat, barley, lentils, peas, chickpeas, ruminants, pigeon, cat and cochineal), China and Southeast Asia (rice, millet, soy, pig, chicken and duck), sub-Saharan Africa (sorghum, rice, millet and faba bean), Mesoamerica (corn, bean, pumpkin and turkey) and the Andes (potato, bean, quinoa and llama). Domestication initiated in one region is generally followed by a migration phase, as human populations travel with their seeds and animals. For example, many plant species were consumed by hunter-gatherers in the Amazon basin. Once domesticated, these species were gradually spread to other regions of the world, following human migrations or, more recently, following the discovery of America, leading to the exploitation in Europe of sunflowers, squash, pumpkins, corn, manioc, taro, potatoes, tomatoes, sweet potatoes, beans, cotton plants and so on.

The motivations behind domestication are certainly diverse. The controlled availability of cereals (and legumes) that could be stored, as well as the possibility of cooking particular foods (wheat bread) or alcoholic beverages (fermented barley) could be, more than a quest for calories, determining factors in the choice of these two main cereals cultivated in the Middle East, whose characteristics are not found in other wild grasses. Participating in the provisioning of social feasts in a region where, at the outset, there was no evidence of demographic pressure, and where fruits and nuts were abundant, a growing cultural interest in these species would have been a determining factor. In the case of animals, three modalities have been proposed to describe domestication (Larson and Fuller 2014). The commensal pathway corresponds to the case of an animal species living in the vicinity of humans with mutual benefit and involves no human intention. The other two pathways involve human intention, with either the predatory pathway, motivated by the need to obtain supplementary food resources (as a possible reaction to over-hunting), or the directed pathway, for any other intention, such as transport or distraction. However, these three pathways are not mutually exclusive: it is likely that domestication first followed a commensal or predatory path before humans acquired the intention of influencing or modifying an animal species for their own benefit.

The processes of domestication and selection have created ever-closer links between agriculture and society, accompanied by changes in human society. Increased yields have made agricultural production systems more dependent on continuous investment in labor, leading to a form of servitude. The domestication of the horse had a major impact on the mobility of human populations, while that of cattle had a major impact on work capacity in the fields.

The genetics of domestication

Which species have been domesticated? How can we identify wild ancestors? What specific morphological or physiological properties characterize it? What genes and gene modifications are involved? Have domesticated organisms been particularly isolated or have they maintained genetic exchanges with their wild ancestor (Hunter 2018)?

Analysis of the ***phylogeny*** of animal and plant species does not show a random distribution of domesticated species. This is particularly clear in the case of animals, where it can be seen that domesticated mammals, known as "cash crops", are derived from only a small fraction of the ***phylogenetic*** and phenotypic diversity of mammals, with an over-representation of the order Artiodactyla (***Bovidae***: cattle, goats, sheep; ***Suidae***: pigs, etc.), and a few species from the orders Rodentia (guinea pig), Lagomorpha (rabbit) and Perissodactyla (horse, donkey). The other domestic species are essentially from the order Carnivora, phylogenetically close to Artiodactyla and Perissodactyla (in the Ferungulata group). Some authors have jointly analyzed the phylogenetic position and biological or ecological particularities of domesticated species, whether they be animal or plant, and have concluded that certain species are predisposed to others (Milla et al. 2018). The domestication of different cereals (wheat, barley, rice, corn, sorghum and millet), characterized by the loss of spontaneous grain dispersal and dormancy, as well as an increase in grain size, has given rise to the concept of a "***domestication syndrome***": common characteristics based on a limited number of mutations, but not implying an identity of these mutations. This concept is only valid if we also accept that characteristics vary according to species type and use: the domestication of cabbage and turnip has led to many different types for the consumption of leaves, roots, stems, inflorescences or the production of edible oil, or for lighting.

Advances in molecular biology and bioinformatics have given us detailed knowledge of the genome that has revolutionized the answers to the questions above. In particular, the study of ancient DNA makes it possible to attribute traits to archaeological remains that do not appear through morphology or chemical composition, such as coat color or fecundity, when the corresponding genetic markers are available (Frantz et al. 2020; Przelomska et al. 2020). It is now possible to compare ancient and recent genomes from domesticated and wild individuals sampled in different parts of the world, in order to estimate the size of the population involved in the initial domestication process, to pinpoint the place or places where a species was domesticated, to identify the genes or chromosomal regions modified by this process, or to retrace the history of populations after domestication. A significant challenge is to be able to distinguish genetic variants that have been favored by domestication, from those that have evolved randomly as a result of genetic drift, or indeed from those whose frequency has changed in response to recent selection. In the wake of domestication, species have been structured into animal breeds or plant varieties, but they may still have exchanges with their wild ancestors. Genetic methods and, where possible, ancient DNA enable us to sort out all the exchanges that have taken place over time.

Domestication is therefore a fascinating process, which has given rise to research in the fields of archaeology, sociology, ecology and genetics. The aim of this book is to illustrate the contribution of the most modern methods of genetic and genomic analysis to our historical understanding of domestication and its biological nature and diversity, based on the examples of 12 species or groups of species from the animal, plant and microorganism worlds, discussed by world-renowned specialists.

References

Frantz, L.A., Bradley, D.G., Larson, G., Orlando, L. (2020). Animal domestication in the era of ancient genomics. *Nature Reviews Genetics*, 21(8), 449–460. doi: 10.1038/s41576-020-0225-0.

Hunter, P. (2018). The genetics of domestication. *EMBO Reports*, 19(2), 201–205. doi: 10.15252/embr.201745664.

Larson, G. and Fuller, D.Q. (2014). The evolution of animal domestication. *Annual Review of Ecology, Evolution, and Systematics*, 45, 115–136. doi: 10.1146/annurev-ecolsys-110512-135813.

Milla, R., Bastida, J.M., Turcotte, M.M., Jones, G., Violle, C., Osborne, C.P., Chacón-Labella, J., Sosinski, E.E., Jr., Kattge, J., Laughlin, D.C. et al. (2018). Phylogenetic patterns and phenotypic profiles of the species of plants and mammals farmed for food. *Nature Ecology & Evolution*, 2, 1808–1817. doi: 10.1038/s41559-018-0690-4.

Przelomska, N.A., Armstrong, C.G., Kistler, L. (2020). Ancient plant DNA as a window into the cultural heritage and biodiversity of our food system. *Frontiers in Ecology and Evolution*, 8, 74. doi: 10.3389/fevo.2020.00074.

Purugganan, M.D. and Fuller, D.Q. (2009). The nature of selection during plant domestication. *Nature*, 457(7231), 843–848. doi: 10.1038/nature07895.

Vigne, J.D. (2015). Early domestication and farming: What should we know or do for a better understanding. *Anthropozoologica*, 50(2), 123–150. doi: 10.5252/az2015n2a5.

Willcox, G. and Stordeur, D. (2012). Large-scale cereal processing before domestication during the tenth millennium cal BC in northern Syria. *Antiquity*, 86, 99–114 [Online]. Available at: http://antiquity.ac.uk/ant/086/ant0860099.htm.

1

Thousands of Years of Relationship between Man and Dog Revealed by Genomics

Morgane OLLIVIER
ECOBIO, Université de Rennes, France

The dog was the first animal to be domesticated by humans, at least 25,000 years ago. Its domestication was made possible by the wide and abundant distribution of its wild ancestor, the wolf, in the northern hemisphere, where human societies were evolving at the end of the Paleolithic era. The wolf's history is therefore atypical of other domesticated species, as its domestication began well before the Neolithic transition. It is not easy to pinpoint a precise date for domestication.

It is important to remember that domestication is a gradual process, and so the dog/wolf distinction is only possible when the characteristics of domestication are visible. The first dogs were probably morphologically very similar to wolves, and so cannot easily be distinguished from one another. Wolves also had a very wide distribution (the entire northern hemisphere), making it all the more difficult to classify remains solely on the basis of geography. Canid fossils are also rare, and consequently the archive is both temporally and geographically fragmented. In this chapter, we will review the main lines of research carried out over the last 20 years, which have shed light on the evolutionary history of the dog and its relationship with man.

Genetics of Domestications,
coordinated by Georges PELLETIER. © ISTE Ltd 2024.

1.1. The history of dog domestication, the subject of much debate

1.1.1. *The dog: domesticated since the Paleolithic period*

The appearance of dogs in the fossil record is recent, and current debates center on whether dogs first appeared in the late Paleolithic or early Neolithic periods. On archaeological grounds, the divergence between dog and wolf is thought to have occurred between 16,000 and 11,000 years BP, a time interval that corresponds to the age of the first undisputed ancient dog remains found in both Europe and Asia at numerous remote archaeological sites (Vigne 2005). Earlier dates have been proposed on the basis of remains initially assumed to be dogs. In Europe, these include remains from the sites of Předmostí in the Czech Republic, estimated at 27,000 years BP, and Goyet in Belgium, estimated at 36,000 years. In Asia, remains found in the Altai Mountains in Russia date back some 33,000 years. However, recent work has cast doubt on these conclusions, demonstrating that the morphology of the specimens studied was within the range of morphological variation of late Pleistocene wolves. For example, fossil Belgian canids, including the Goyet dog, form a sister group to all contemporary dogs and wolves, suggesting that they represent either a failed domestication event, or a morphologically distinct form of wolf.

In 2015, the genome of a 35,000-year-old wolf from the Taimyr Peninsula in northern Siberia was sequenced. By comparing this genome with that of the modern wolf and that of the present-day dog and based on an estimate of the mutation rate between these genomes, it was concluded that the ancestors of dogs were separated from present-day wolves before the Last Glacial Maximum and the Taymir wolf (Figure 1.1) belonged to a population that had already diverged from the common ancestor of modern wolves and dogs. Dogs are probably descended from extinct wolf populations that diverged from the ancestors of extant wolves around 27,000–40,000 years ago (Skoglund 2015). Present-day gray wolves would not be their direct ancestors, and it is likely that the modern-day dogs, which are closest to these wolves, are the result of multiple hybridizations (Frantz et al. 2016). The disappearance of these populations could be explained by the decline in wolf populations during the Last Glacial Maximum, which then led to the repopulation, expansion and replacement of ancestral populations by current populations. The wolf population from which dogs most likely descend were large northern wolves, and none of the current wolf lineages from the hypothetical regions of domestication appear to be the original source population (Skoglund et al. 2015). These studies therefore suggest that dogs probably originated in Eurasia as early as 33,000 years ago, an earlier date than their first record in the archaeological record, and that their evolutionary history is complex along a long phenotypic continuum.

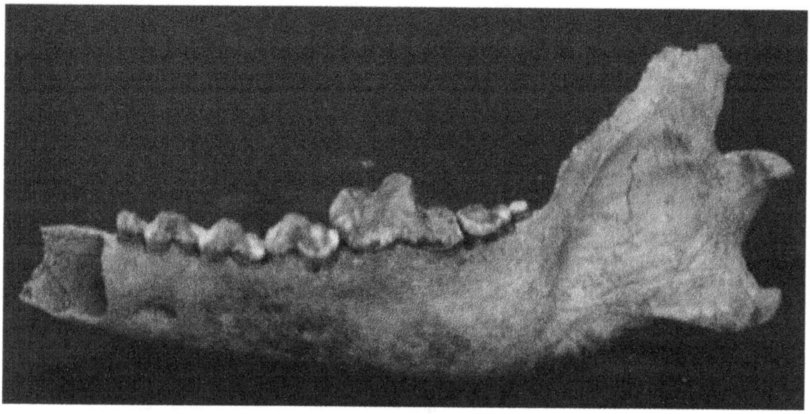

Figure 1.1. *Taymir wolf mandible fragment (source: Reuters). For a color version of this figure, see www.iste.co.uk/pelletier/genetics.zip*

1.1.2. *The process behind dog domestication*

Domestication leads to the appropriation and control by a human society of an animal or plant sub-population for the production of a service or commodity. It involves an evolutionary process that documents two aspects of evolution: speciation and adaptation. In the domestication process, these two biological mechanisms consist, on the one hand, of a relative break in gene flow between the original population and the domesticated population and, on the other hand, of a modification in selective pressures. Based on ethnographic research suggesting that domestic animal husbandry was not unusual among hunter-gatherer groups worldwide, Francis Galton suggested that dogs had been domesticated following the capture and rearing of cubs in human camps. Later, Belyaev (1969) showed that animal domestication could have begun in the absence of deliberate human design and action. In 2011, Vigne (2011) proposed a multi-stage model characterized by a gradually intensifying relationship between humans and animals. In this perspective, animal domestication proceeded along a continuum from anthropophilia to commensalism, control in the wild, control of animals in captivity, extensive breeding, intensive breeding and, finally, pets. While Zeder (2012) has also acknowledged this model's step-by-step approach, she describes three distinct paths that animals may have followed in a domestic relationship with humans: a commensal path, a prey path and a directed path.

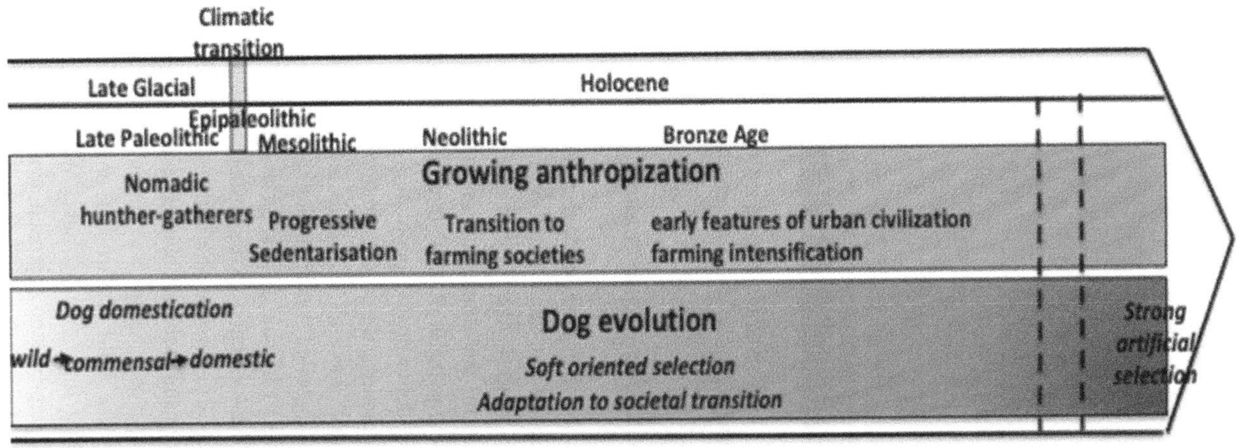

Figure 1.2. *Schematic representation of the chronology of human societal change and the history of the dog in Western Europe (source: M. Ollivier). For a color version of this figure, see www.iste.co.uk/pelletier/genetics.zip*

In the case of the dog, it is assumed that the domestication process first took the form of commensalism/mutualism (Vigne 2011) between wolves and humans at the end of the Paleolithic era (Vigne 2005). The commensal pathway does not begin with intentional human action to bring wild animals (juvenile or otherwise) into their camps. Instead, different wild animal populations would have been attracted by elements of the human niche, including human food waste.

The individuals most able to take advantage of the resources associated with human camps would have been the more docile, less aggressive individuals. A recent study supported this view, proposing a hypothesis based on the sharing of resources between humans and wolves. Humans are not fully adapted to a carnivore diet, as their meat consumption is limited by the liver's ability to remove the ammonia produced by protein digestion.

Unlike humans, wolves can thrive on lean meat for months on end. Yet the study shows that all Pleistocene archaeological sites where dog remains have been found originate from areas similar to subarctic and arctic environments. Calculations show that during harsh winters, when game was lean and lacking in fat, late Pleistocene hunter-gatherers in Eurasia would have benefited from a surplus of animal protein that could have been shared with early dogs. This study explains how competition may have contributed to the initial phase of domestication.

Thus, the relationship between man and dog probably involved a gradual taming of the wolf linked to a cultural process (Larson and Burger 2013), particularly in Europe, where it is likely that such interactions began over 30,000 years ago. After the first phase of domestication, dogs would have become more docile, being used in a multitude of ways, such as hunting companions, beasts of burden, guards, and would have undergone many evolutionary changes similar to those of humans.

1.1.3. *The number and location of domestication events: contextualization and the contribution of archaeological and genomic data*

Gray wolves were distributed throughout the northern hemisphere, and it is assumed that the wolf populations that gave rise to dogs, in whole or in part, are certainly extinct. Furthermore, the human populations that domesticated wolves were mobile hunter-gatherers. For these reasons, it has long been difficult to establish where dogs were domesticated, or even whether several wolf populations were domesticated independently. Several morphological groups can also be distinguished among Upper Paleolithic dogs: large, robust canids are observed in northeastern and central Europe, while medium-sized dogs with strong allometric differences are described in the Middle East. Lastly, small-sized individuals are found in southwestern Europe. This morphological diversity, as well as the temporal concordance of dog samples found in both Europe and Asia, with the wide distribution of wild ancestors and Upper Paleolithic dogs, has been interpreted by archaeozoologists since the 19th century as a strong argument for multiple, independent domestications.

Genetic studies of the diversity of today's dog breeds, on the other hand, have led to contradictory and sometimes opposing conclusions. Analysis of the ***mitochondrial DNA*** of 162 wolves from 27 locations worldwide, and 140 domestic dogs representing 67 breeds, shows that most dog sequences belong to a single group in which no wolf sequences are found. However, in three other groups, both dog and wolf mitochondrial sequences are found, which the authors interpret as an initial domestication event, followed by repeated cycles of hybridization and selection for canine phenotypic variation. Repeated genetic exchanges between dog and wolf populations may have been a significant source of variation for artificial breed selection. Analysis of a 582-base-pair region of mitochondrial DNA from 654 domestic dogs from Europe, Asia, Africa and Arctic America, as well as 38 Eurasian wolves, shows that haplogroup A1 is dominant over the other five (B, C, D, E and F), and that dogs with a Southeast Asian origin show greater genetic diversity. This DNA region was also analyzed in 1543 Old World dogs, including modern breeds of known geographical origin, as well as breeding dogs of indigenous ancestry, which are more relevant for unravelling the origins of the dog than modern breeds. The results point in the same direction, showing greater genetic diversity in Asia, with Western Eurasian ***haplotypes*** all attributed to an East Asian haplotype. These studies thus proposed East Asia as the main and only center of domestication, based on comparisons of genetic diversity between regions, hypothesizing that centers of origin contain greater genetic variation. A new interpretation was subsequently given to this observation by some authors, who suggested that it reflected a replacement of native European canine ancestry by more recent lines of Southeast Asian immigrants. This replacement may have been accompanied by a general geographical expansion of East Asian dog populations or, alternatively, explained by human transportation in connection with trade.

Other studies, again based on modern DNA, suggest, on the contrary, and in line with archaeological evidence, that several centers of domestication existed in Eurasia. For example, mitochondrial DNA diversity in African village dogs is comparable to that of East Asian dogs. This discovery was used to question the Southeast Asian origin of the dogs. Subsequent analysis of 48,036 nuclear genome ***SNP*** positions in 912 modern dog breeds, 225 wolves and 60 coyotes, demonstrates that Middle Eastern wolves contribute a significant amount of genomic diversity to modern dog breeds, suggesting a Middle Eastern contribution to the history of dog domestication.

Various observations may explain why it is so difficult to unravel the origins of dogs from current genetic data: (i) most dog breeds have very recent origins, with a history going back less than 150 years; (ii) dog populations have undergone numerous episodes of diversification, hybridization and homogenization, reducing the resolving power of genetic data from modern breeds; (iii) a combination of ***introgressions*** and bottlenecks between dog and wolf populations, well after the domestication process, may have blurred modern genomic signatures and reduced our ability to make a distinction between the initial domestication process(es) and post-domestication ***gene flow***; (iv) the wolf population(s) that gave rise to modern dogs may be extinct, making it difficult to detect the initial process using studies based on modern data. The ***paleogenomic*** approach therefore appears to be the most effective tool for tracing the history of dog domestication.

1.2. Paleogenomics: an essential tool for understanding the evolutionary history of the dog

Understanding the origins and early history of dogs has been complicated by bottlenecks, expansions, extinctions as well as local replacements and gene flows between wolves and dogs. Large-scale systematic research and analysis of ancient wolf and dog genomes across space and time is therefore needed to accurately reconstruct the evolutionary history of the first domestic animal.

1.2.1. *Eurasian origins and diffusion*

In 2013, the first mitochondrial genomes of ancient dogs and wolves were published and began to provide answers. The results show that dogs and wolves do not form two distinct ***monophyletic groups***. Each of the ***clades*** grouping modern dogs of haplogroups A, C and D have an ancient lineage as a sister group, grouping ancient European dogs and wolves. It was thus demonstrated that dogs have a European origin and have been descended from a line of wolves for over 20,000 years, a line that is now extinct. Following this, the analysis of over 99 mitochondrial sequences of Upper Paleolithicto Bronze Age individuals from 29 archaeological sites in Eurasia, and the complete genome of the Newgrange dog (late Neolithic, ~ 4,800 ago) in Ireland suggested that dogs, in both Eastern and Western Eurasia, were independently domesticated in the Upper Paleolithic from several wolf populations (Frantz et al. 2016). Later, during the Neolithic and early Bronze Age, Asian dog lineages spread westwards, leading to a profound change in the European dog population (Ollivier et al. 2018). Two types of diffusion seem to have taken place: one linked to the Neolithic transition, which may have originated in the Middle East, and a later one from Asia to Europe at the end of the Neolithic/Chalcolithic period, probably via the Pontic steppes and accompanied by other elements such as millet, the domestic horse and early metallurgy. Thus, human movements and cultural diffusion have profoundly shaped the lineage of dogs. Lastly, a very recent study shows that this scenario could be even more complex. By sequencing 27 ancient dog genomes from across Eurasia, Bergström et al. (2020) confirm that all dogs derive from a lineage of wolves that are distinct from present-day wolves and also show, as suggested by archaeozoological data, that five major lineages were already present and diversified 11,000 years ago, as early as the Paleolithic period. The fate of each of these lineages is therefore linked to, and even mirrors, migrations and the history of human populations. By studying the evolutionary history of the dog, we can retrace the history of mankind.

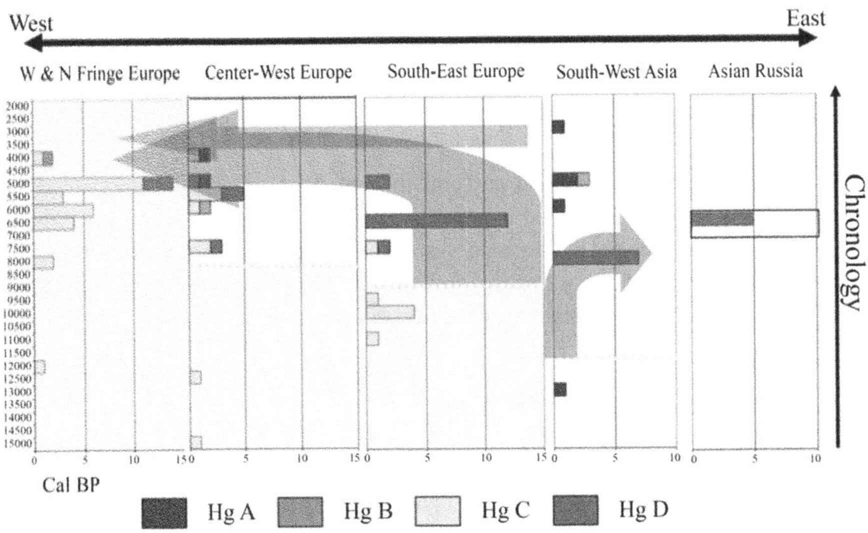

Figure 1.3. *Dog population movements in Europe between the Paleolithicera and the Bronze Age, inferred by mitochondrial data (Hg: haplogroup; BP: before present) (source: adapted from Ollivier et al. (2018)). For a color version of this figure, see www.iste.co.uk/pelletier/genetics.zip*

1.2.2. *The uniqueness of the Iberian Peninsula*

The Iberian Peninsula, like Italy and the Balkans, was a refuge during the Ice Age, preserving a unique diversity. The earliest evidence of dogs dates from

7,900 to 7,600 years ago in archaeological contexts accumulated by hunter-gatherer societies, a few centuries before the arrival of the first Neolithic communities.

By analyzing the remains of five Mesolithic dogs and 15 samples of dogs dating from the Roman period, collected from archaeological sites in Portugal, the authors show a high frequency of mitochondrial haplogroup A, the majority in present-day dogs. Eighty percent of the Mesolithic sequences from Portugal belong to this haplogroup. This result provides evidence that haplogroup A was present in Europe before the Neolithic period. Until now, haplogroup C was considered the most common clade in Europe before the Neolithic era (Frantz et al. 2016). Haplogroup A, observed at higher frequencies in the Middle East and Asia, had been considered to have been introduced to Europe after the arrival of Neolithic farmers, perhaps in the Bronze Age (Ollivier et al. 2018). These results suggest, in contrast, that this haplogroup may have been present in European refuges prior to Neolithic influence. These Mesolithic dogs could be attributed to local domestication(s), which is a scenario corroborated by mitochondrial data from Paleolithic Iberian wolves. Among the samples dating from the Roman period, 12 from Portugal, Spain and Morocco belong to clade A, and three from Spain belong to clade D. So far, this is the oldest evidence of clade D in the Iberian Peninsula. This result supports the idea that there may have been significant consolidation of dog breeds from divergent genetic lineages during this period. According to these data, local breeding involving dogs of clade A and clade D in the Iberian Peninsula was continuous for at least 1,600 years. In addition, the sharing of lineages between Spanish and North African dogs may suggest genetic flow. Dogs could have been easily transported between these regions by humans following shipping trade routes.

1.2.3. *Origins of the dog in America*

The oldest dog records in North America are found in Alaska and Illinois, dating back 9,020 ± 85, 8,820 ± 30 and 8,840 ± 80 years, respectively (Perri et al. 2019; da Silva Coelho et al. 2021). The origins and fate of dog populations in the Americas, prior to contact with European and African peoples, have been the subject of recent study involving comparisons of ancient and modern dog genomes. Comparative genomic analyses demonstrated that the first American dogs had no ancestry with American wolves. On the contrary, these Pre-Contact Dogs (PCDs) represent a specific lineage that may have migrated from Northeast Asia across the Beringian steppe alongside humans, over 10,000 years ago. These analyses also showed that PCD populations were subsequently almost completely replaced by European dogs during the large-scale colonization of North and South America over the last 500 years. The introduction of infectious diseases during this recent colonization probably played a major role in the replacement of PCDs by European dogs.

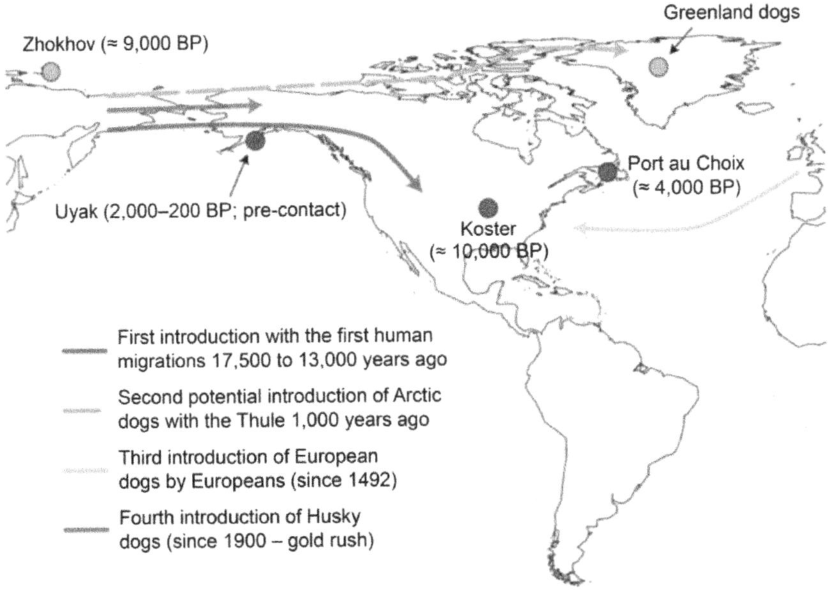

Figure 1.4. *Multiple dog introductions in North America (BP: before present) (source: M. Ollivier, adapted from Leathoblair et al. (2019)). For a color version of this figure, see www.iste.co.uk/pelletier/genetics.zip*

The introduction of dogs into South America remains to be clarified, but most probably follows the introduction of dogs into North America via Beringia, after the arrival of humans in South America. Indeed, the oldest dog remains found in America are in the north of the continent and post-date the arrival of humans in the far south of the continent 11,000 years ago. Analysis of the mitochondrial genomes of North American dogs has shown that pre-Columbian dogs belong to a monophyletic group (haplogroup A), having diverged 14,600 years ago, sharing a common ancestor with East Siberian dogs at least 16,700 years ago. The most likely hypothesis is that they migrated with human populations from North to South America. Mitochondrial DNA analysis of several pre-Columbian Mexican dog specimens (Manin et al. 2018) supports this hypothesis. Four haplotypes belonging to haplogroup A were identified. Three of these haplotypes are common to dogs of European and American origin. These dogs, probably originating from Eurasian populations that colonized North America, reached Central and South America, where they remained isolated for several thousand years. The last haplotype, however, is close to a sequence that is only shared by modern-day Chihuahua dogs, which is thought to have originated in Mexico. This proximity suggests that it is a characteristic haplotype of pre-Columbian dogs, resulting from the genetic divergence of these local populations over the thousands of years following their introduction. Lastly, a recent ***phylogenetic*** analysis of ancient mitochondrial genomes from South American dogs confirms that pre-Columbian and modern South American dogs share a common ancestry and belong to haplogroup A. In this study, the authors also show that present-day American dogs and pre-Columbian dogs show a high degree of genetic divergence, supporting the hypothesis that pre-Columbian dogs were almost completely replaced by European dog populations after the arrival of Christopher Columbus.

1.3. From commensalism to modern breeds: identifying the genomic foundations behind the intensification of the human–dog relationship

Domestication involves an evolutionary process that documents two aspects of evolution: speciation and adaptation. In the domestication process, these two biological mechanisms consist, on the one hand, of a relative break in gene flow between the original population and the domesticated one, and, on the other hand, of a modification in selective pressures. Some studies have highlighted the importance of progressive intensification of the relationship between man and animals, leading, in the case of the dog, to a continuum from commensalism, to captive animal control, to breeding and, ultimately, to pets (Vigne 2011; Zeder 2012). Isolating, taming, controlling and moving animals into an anthropogenic ecosystem has morphological, physiological and behavioral consequences that can be identified as a sequence of common traits selected along this trajectory from wild to domesticated (Belyaev 1969; Arbuckle 2005, pp. 18–33; Trut et al. 2009). During the process of domestication, a relaxation of natural selective pressures allowed the persistence of mutations linked to these new traits, resulting in an ever-increasing divergence between the genomes of domesticated animals and those of their wild ancestors. Unlike other domesticated animals, domesticated canids have also been subject to progressive human impact on their environment, as a result of changes in socioeconomic systems and human subsistence patterns. Such shifts in cultural contexts, as well as the resulting new selective pressures, could also have had an impact on the evolution of the dog genome, by leveraging changes already in place, or enabling the emergence of new changes. In general, however, it is difficult to identify regions of the genome that are unambiguously associated with early stages of domestication, and then implicated in current phenotypic diversity. This is because (i) there may only be a very limited number of regions involved in this process, a phenomenon particularly highlighted in dogs; (ii) some traits (such as behavior) have a complex genetic basis such as multiple gene inheritance and epistatic relationships[1], or are linked to regulation of gene expression; (iii) the regions identified may be large, without it being possible to accurately identify the genes or variants involved; (iv) studies are costly and time consuming (Larson and Burger 2013; Larson et al. 2014). Despite this, the change in scale of analysis (from the candidate gene to the genome) in recent years has made it possible to study the domestication process at high resolution in order to determine how it has shaped modern domesticated animals. However, understanding the basis and genetic architecture of the domestication process remains a challenge.

1.3.1. *Selection and domestication markers*

The mitochondrial and nuclear genomes of the dog possess an overabundance of ***non-synonymous mutations*** when compared to wolves and coyotes, which can be explained by a relaxation of selective pressures after domestication. Models suggest that once animals are freed from the selective pressures associated with life in the wild, they accumulate non-synonymous mutations under a relaxed selection regime, some of which led to new phenotypes that humans might have preferentially selected (consciously or unconsciously), while nature would have actively eliminated them. This increase in the number of non-synonymous mutations is observed in all modern dog breeds, suggesting the presence of small populations during and after domestication.

The detection of genomic ***loci*** with ***selective sweep*** signatures allows us to identify candidate regions that may have undergone rapid micro-evolution during, and shortly after, domestication. Many of the regions identified concern lipid or starch metabolism, immunity, behavior, brain function and pigmentation. These findings are consistent with the shared history of humans and domestic dogs. For example, with the development of agriculture, metabolic mechanisms to digest increasing proportions of starch from cereal crops were selected. Thus, among the candidate genes for the domestication process, the *MGAM* gene has been identified as being under positive selection. This gene is involved in the final stages of starch digestion. Similarly, one of the strongest selection signals between dogs and wolves is centered around *CCRN4L*, a gene that controls lipid metabolism through its interaction with *PPAR-Y*. The selection of this candidate gene is consistent with the change in lipid content in the diet of the earliest domestic dogs. When dogs and humans began to hunt together, the number of prey caught probably increased

1 Functional interactions between different genes, like when an allele of gene A masks the effect of an allele of gene B.

relative to the number caught by wolves alone, and with it, the amount of lipid consumed by early dogs. A unique dietary selective pressure may thus have resulted from both the amount of lipids consumed and the composition of the tissues made available to these dogs after humans had removed the best parts of the carcass. The dogs' behavior was also markedly altered from that of the wolves in terms of docility and canine understanding of human facial expressions and actions. Thus, all comparisons of dog and wolf genomes revealed evidence of selection on candidate neurobehavioral genes. For example, *MBP*, a schizophrenia-related gene responsible for myelin sheath formation, and *SH3GL2*, a gene involved in synaptic vesicle formation, were found in regions identified as being under positive selection in dogs (Axelsson et al. 2013).

1.3.2. *Genetic variation and adaptation to a starch-rich diet*

In western Eurasia, the Neolithic transition took place between 11,500 and 6,000 years ago, with the transition from hunting and gathering to agriculture. By this time, the dog, which was domesticated during the Upper Paleolithic, had already been accompanying humans for several millennia. The study of the impact of changes in diet, which was brought about by these human societies, on the physiological changes of the first domesticated canids and the transformations in their genomes that gave rise to today's canine genotypes and phenotypes, is of crucial importance.

A comparison of current dog and wolf genomes has identified genomic regions potentially affected by this process. In particular, a series of duplications of the gene coding for pancreatic amylase (*Amy2B*) appears to have been selected, leading to an increase in the number of copies of this gene in modern dogs. An increased number of *Amy2B* copies is associated with higher amylase activity, enabling improved starch digestion capacity (Axelsson et al. 2013). While the number of *Amy2B* copies widely varies in dogs (4–34 copies), the number is much lower (two to eight copies) in wolves, with 60% of wolves carrying only two copies of the gene. This suggests that dogs are better adapted to a starch-rich diet than wolves. Present-day canids show three patterns of variation in the number of *Amy2B* copies: (i) 60% of wolves and most dingoes carry two copies of the gene; (ii) dogs and wolves carrying between two and eight copies of *Amy2B*; (iii) dogs carrying more than eight copies of *Amy2B*. However, the question of a link between the increase in the number of *Amy2B* copies in dogs and the Neolithic transition remains unanswered with these data, because while this increase could have provided a strong adaptive advantage in an agricultural context, we cannot rule out that it occurred much later, following the more recent selection of specialized lineages.

Paleogenetics offers a unique opportunity to shed light on this question, through studying the genetic landscape of the varying number of *Amy2B* copies in ancient canid populations. The study of the number of copies of the *Amy2B* gene in 88 ancient dogs from various archaeological sites across Eurasia yielded results for 13 of them. Four of the ancient dogs – from Romania (6,500 years ago), Turkmenistan and France (4,000–5,000 years ago) – possessed more than eight copies of *Amy2B*. The hypothesis of a modern origin explaining the increase in the number of copies is ruled out. The earliest expansion of the *Amy2B* gene number could be dated to 7,000 years ago (Neolithic era) and probably constituted an important adaptive advantage for dogs that were able to feed on the waste or remains of human meals (Ollivier et al. 2016). Interestingly, five of the samples appear to only possess two copies, which is a situation that is almost nonexistent in present-day dogs. Dog populations with low *Amy2B* copy numbers are currently associated with nonagricultural populations (Arent 2016). This pattern supports the hypothesis of an ancient expansion of *Amy2B* associated with the Neolithic period. Ultimately, it is worth noting that individuals presenting the three profiles (two copies, three to eight copies, more than eight copies) can be found on the same archaeological sites, which shows that in the Neolithic era, the expansion of the *Amy2B* gene was not yet fixed in dog populations associated with Neolithic agricultural societies. These results have recently been corroborated (Bergstrom et al. 2020) by showing that several Neolithic dogs (Iran: 5,800 years old; Spain: 6,200 years old) do indeed possess a high number of copies of the *Amy2B* gene. This increase, which was variable among dog populations associated with the first farmers, was widespread and fixed for a few thousand years with the regular use of starch-rich agricultural products.

1.3.3. *The evolution of coat color*

Analysis of gene sequences encoding phenotypic traits in ancient dogs can provide information on their phenotype, on the diversity of the trait in ancient populations, and may even be an indicator of the impact of domestication. Changes in coat color is often considered one of the traits that can vary from earliest stages of domestication (Belyaev et al. 1969; Trut et al. 2009). Genetic landscape analysis of coat color variation has shown the presence of such variations in ancient populations of Eurasian dogs and wolves. The isolation of DNA fragments of two genes controlling coat color, *Mc1r* (melanocortin 1 receptor) and *CBD103* (canine-β-defensin), from 14 different archaeological sites spanning a period from the late Upper Paleolithic to the Bronze Age, detected a new variant (*R301C*) of the melanocortin 1 receptor, highlighting the presence of the melanistic beta-defensin mutation (*CDB103-K* locus) in the genomes of several ancient dogs. The results show that the dominant

K^B (*CBD103*) allele, which is responsible for melanism, and the *R301C* (*Mc1r*) variant, which can cause light coat coloration, have been present since the beginning of the ***Holocene*** era in Eurasia, over 10,000 years ago. These results underline the genetic diversity of prehistoric dogs, inherited not only from the wild pool from which the dogs originated (*R301C* variant), whose variability remains largely unknown, but also from mutations most likely linked to the relaxation of natural selection (such as the K^B variant).

1.3.4. *The question of human–dog co-evolution*

Multiple studies aimed at understanding the origin and spread of dogs around the world have highlighted that dogs have accompanied humans throughout their migrations (Ollivier et al. 2018; Bergström et al. 2020). Joint analysis of ancient human and canine genomes has also revealed that part of the history of canine populations mirrors that of humans (Bergström et al. 2020). Thus, the diversification of at least five lineages of canine ancestry, at the beginning of the Holocene era, was probably the result of a dynamic reflecting the way dogs migrated alongside humans. However, this phenomenon is not a generality and, in many cases, the geographical distributions of dogs and humans do not always correspond, revealing a complex history of different canine lineages. This suggests that humans have also migrated without dogs, that dogs have moved between human groups or, like other domestic animals, dogs may have been the object of commercial, cultural or economic exchanges.

An example of bio-cultural co-evolution is the increase in the number of copies of the *AMY2B* gene involved in starch digestion, linked to dietary adaptations in dogs during the agricultural transition (Axelsson et al. 2013; Ollivier et al. 2016; Bergström et al. 2020). The homologous[2] human *AMY1* gene has probably undergone adaptive evolution, although this has not been clearly linked to agriculture. It is therefore reasonable to hypothesize on other examples of co-evolution between human culture and canine genomic variants. It is even possible to assume at least partial parallel evolution between the genomes of the two species. To support this hypothesis, a list of genes that have undergone positive selection in the dog's evolutionary history has been compiled, which largely overlaps with the corresponding list of positively selected loci in humans. The most apparent parallel evolution concerns genes involved in digestion, metabolism, neurological processes and cancer. One example of a locus under positive selection concerns a 5 Mb genomic region on chromosome 6. Deletion of the homologous region (carrying the

2 Homologous genes are present in different species that have evolved from the same ancestral gene.

GTF2I and *GTF2IRD1* genes) in humans is linked to Williams–Beuren syndrome (WBS), a multi-systemic, disabling congenital disorder, one component of which is hyper-social behavior. However, hyper-sociability is also a central feature of domestication that distinguishes dogs from wolves. Structural variation in the same region in dogs also contributes to extreme sociability. This discovery suggests that there are commonalities between the genomic architecture of the WBS and canine taming, and that directional selection may have targeted a unique set of genes enabling rapid behavioral divergence of dogs and wolves, facilitating coexistence with humans.

A bond involving gaze also exists between humans and dogs. Mutual gaze is thought to increase the release of oxytocin, which in turn increases their propensity to stare at their master, which also has an effect on humans. In contrast, wolves, which rarely make eye contact with humans, seem resistant to this effect. Humans and dogs are locked in an oxytocin feedback loop that is partly mediated by this sustained eye contact.

We can give one last example of parallel evolution. In the same way that the variant of the *EPAS1* gene, inherited from the Denisovan man giving adaptation to life at altitude, was selected in *Homo sapiens* and is found in Tibetan human populations, the *EPAS1* homologue, derived from altitude-adapted gray wolves and has also been selected in the Great Danes that have populated the Tibetan plateau for several hundreds of years.

Understanding how humans and dogs interact is therefore important, as specific human practices and mutual cooperation between the species can explain changes in the physical, behavioral or functional appearance of dogs.

1.4. Selection of modern breeds, evolution in an anthropic context

1.4.1. *Origins and timing of selections*

Today, dogs have many different roles, ranging from protection, hunting and herding to companionship, revealing very different selective pressures over time. Today, in Europe, dogs have no economic value for food production and are mainly companions for humans. With 85.2 million individuals in Europe (25% of families have at least one dog in France or the UK; 42% in Poland or Romania), dogs play a major social role. The process of domestication was one of the first causes of morphological changes at the origin of the current phenotypic diversity of dogs (in addition to those from the wild pool). This has been demonstrated in fox domestication experiments (Trut et al. 2009) as a model (loss of wild color, drooping ears, shortening of limbs, curved tail, wider muzzle, reduction in size, etc.)

and by the archaeological approach (widening of the skull, shortening of the mandible or overlapping of teeth). Since their domestication, the dog's genome has evolved as a result of (i) unconscious or directed human selection for preferential phenotypes, or phenotypes that facilitate domestication; and (ii) the adaptation of animals to variations in their new environment and to their new lifecycles, imposed by chronological and cultural contexts. In the present state of knowledge, the timing and speed of the first human-mediated selections, as well as the traits selected by humans, remain unknown. However, it is possible that voluntary selection of phenotypic traits by humans began early in the dog's evolutionary history. This selection could have been initiated by controlling the reproduction of individuals carrying preferentially selected traits that were well adapted to local conditions and could have been highly dependent on socio-cultural contexts. For example, at the Ust-Polui site (Arctic, Bronze Age), the fact that some dogs were ritually buried, while others were slaughtered, suggests a complex set of beliefs about the dog's place in society. Canines were not considered a homogeneous group at this site. These findings suggest that humans built lasting relationships with individuals they deemed valuable to the community, molding them to their needs and cultural specificities. These diverse practices – whether intentional or not – may therefore have influenced the evolution of the dogs, shaping the physical characteristics and personality traits of the animals that lived alongside these communities.

The emergence of the concept of breed standard, some 300 years ago, eventually led to the exacerbated artificial selection of certain phenotypic traits and the fragmentation of the original domestic populations into over 350 breeds. At the root of this phenotypic variability, analyses of modern dog genomes have revealed a handful of mutations (not found in existing wolves) with significant effects on morphological variation (Plassais et al. 2019), such as coat characteristics, length, size, weight and height variation, ear morphology, coat color and behavior (a nonexhaustive list). As this artificial selection is very recent, there is no way of deducing the phenotypes of ancient dogs and the variability of ancestral genotypes from modern data, as most of the initial genetic and phenotypic variation may have been lost in this recent process.

1.4.2. *Phenotypic and genetic variability*

Understanding and identifying the genetic basis of phenotypic variability in dogs has long been the main focus of genetic, and later genomic, studies. Much of the dog's phenotypic diversity derives from variants that are not present in the wolf, which therefore emerged after domestication. Classical genetic analyses have succeeded in identifying certain genes that cause genetic diseases and discrete phenotypic traits, such as coat color variation. However, the identification of the genetic architecture of complex traits, such as body size and conformation, has only been possible with the development of ***GWAS (genome-wide association study) analyses***. These studies have shown that in dogs, most of the variation in a phenotypic trait is due to variation in just a few genes. This probably reflects the method of artificial selection used to fix specific traits, involving few breeding animals, often from the same litter. In contrast to the mutations responsible for each breed's characteristics, carried by just a few genes, the genetic basis of the divergence between dog and wolf appears to be more complex, involving a polygenic architecture. Few loci in coding areas of the genome appear to be associated with this wolf/dog divergence, suggesting a crucial role for mutations in regulatory areas in the evolution of dog-specific traits. Today, genomic analyses suggest the existence of six to nine main groups of dog breeds. The richness of phenotypic diversity is explained by the incredibly small number of mutations shared between breeds, compared to other domestic species. However, while the dog is the species with the greatest phenotypic variability, with over 350 breeds, genetic diversity within each breed is very low.

1.4.3. *Harmful effects and genetic diseases*

While advantageous genomic regions may have been selected, domestication also has detrimental effects on a genomic scale, as stated at the start of section 3.1. Domestication is associated with multiple bottlenecks, and it is therefore expected that the effectiveness of selection in eliminating harmful mutations will be reduced. This effect is amplified by artificial selection, in the course of which breeds are genetically homogenized, notably through the use of a low number of breeding stock, thereby increasing inbreeding and reducing effective population size.[3] Today, this phenomenon is the main cause of genetic disease in dogs. In dogs, there is a link between the genes involved in Mendelian genetic diseases and the genes located in these selective sweep areas that are responsible for the formation of breeds. This link is significantly higher in dogs than in human populations. This is the case, for example, with selection for merle color, one of the most popular coat patterns in dogs. Mating between merle individuals can result in double merle (MM) offspring with eye defects (including blindness and microphthalmia) and hearing defects. Dog breeders, in selecting phenotypic traits, often unknowingly target mutations present in adjacent regions, resulting in genetic diseases (such as chondrodysplasia,

3 The size of a theoretical population which would evolve, from the point of view of inbreeding and genetic drift, with the same speed as the population studied. This theoretical population is closed, of constant size, made up of individuals who reproduce randomly, without selection or mutation.

dwarfism, hereditary cancers, obstructive brachycephaly and dysplasia). Certain phenotypes found in dogs are observed in wild canid populations, such as chondrodysplastic limbs (shortened limbs similar to those of dachshunds, corgis or basset hounds), but they are however counterselected, given their extreme impact, in this case, on movement. Modifications to breeding programs and guidelines must therefore include avoiding inbreeding and ensuring greater variation within breeds, so as to prevent the accumulation and expression of harmful mutations and improve dog health.

1.5. References

Arbuckle, B.S. (2005). Experimental animal domestication and its application to the study of animal exploitation in prehistory. In *The First Steps of Animal Domestication: New Archaeozoological Approaches*, Vigne, J.D., Peters, J., Helmer, D. (eds). Oxbow, Oxford.

Axelsson, E., Ratnakumar, A., Arendt, M.J., Maqbool, K., Webster, M.T., Perloski, M., Liberg, O., Arnemo, J.M., Hedhammar, A., Lindblad-Toh, K. (2013). The genomic signature of dog domestication reveals adaptation to a starch-rich diet. *Nature*, 495, 360–364. doi: 10.1038/nature11837.

Belyaev, D.K. (1969). Domestication of animals. *Science Journal*, 5, 47–52.

Bergström, A., Frantz, L., Schmidt, R., Ersmark, E., Lebrasseur, O., Girdland-Flink, L., Lin, A.T., Storå, J., Sjögren, K.G., Antony, D. et al. (2020). Origins and genetic legacy of prehistoric dogs. *Science*, 370(6516), 557–563. doi: 10.1126/science.aba9572.

Brown, S.K., Pedersen, N.C., Jafarishorijeh, S., Bannasch, D.L., Ahrens, K.D., Wu, J.T., Okon, M., Sacks, B. (2011). Phylogenetic distinctiveness of Middle Eastern and Southeast Asian village dog Y chromosomes illuminates dog origins. *PLoS ONE*, 6(12). doi: 10.1371/journal.pone.0028496.

Frantz, L.A.F., Mullin, V.E., Pionnier-Capitan, M., Lebrasseur, O., Ollivier, M., Perri, A., Linderholm, A., Mattiangeli, V., Teasdale, M.D., Dimopoulos, E.A. et al. (2016). Genomic and archaeological evidence suggests a dual origin of domestic dogs. *Science*, 352(6290), 1228–1231. doi: 10.1126/science.aaf3161.

Larson, G. and Burger, J. (2013). A population genetics view of animal domestication. *Trends in Genetics*, 29(4), 197–205. doi: 10.1016/j.tig.2013.01.003.

Larson, G., Piperno, D.R., Allaby, R.G., Purugganan, M.D., Anderson, L., Arroyo-Kalin, M., Barton, L., Vigueira, C.C., Denham, T., Dobney, K. et al. (2014). Current perspectives and the future of domestication studies. *Proceedings of the National Academy of Sciences of the United States of America*, 111(17), 6139–6146. doi: 10.1073/pnas.1323964111.

Ollivier, M., Tresset, A., Bastian, F., Lagoutte, L., Axelsson, E., Arendt, M.'L., Balasescu, A., Marshour, M., Sablin, M.V., Salanova, L. et al. (2016). Amy2B copy number variation reveals starch diet adaptations in ancient European dogs. *Royal Society Open Science*, 3(11). doi: 10.1098/rsos.160449.

Ollivier, M., Tresset, A., Frantz, L.A.F., Bréhard, S., Balasescu, A., Mashkour, M., Boroneant, A., Pionnier-Capitan, M., Lebrasseut, O., Arbogast, R.M. et al. (2018). Dogs accompanied humans during the Neolithic expansion into Europe. *Biology Letters*, 14(10). doi: 10.1098/rsbl.2018.0286.

Plassais, J., Kim, J., Davis, B.W., Karyadi, D.M., Hogan, A.N., Harris, A.C., Decker, B., Parker, H.G., Ostrander, E.A. (2019). Whole genome sequencing of canids reveals genomic regions under selection and variants influencing morphology. *Nature Communications*, 10(1). doi: 10.1038/s41467-019-09373-w.

Skoglund, P., Ersmark, E., Palkopoulou, E., Dalén, L. (2015). Ancient wolf genome reveals an early divergence of domestic dog ancestors and admixture into high-latitude breeds. *Current Biology*, 25(11), 1515–1519. doi: 10.1016/j.cub.2015.04.019.

Trut, L., Oskina, I., Kharlamova, A. (2009). Animal evolution during domestication: The domesticated fox as a model. *BioEssays*, 31(3), 349–360. doi: 10.1002/bies.200800070.

Vigne, J.D. (2011). The origins of animal domestication and husbandry: A major change in the history of humanity and the biosphere. *Comptes Rendus Biologies*, 334(3), 171–181. doi: 10.1016/j.crvi.2010.12.009.

Vigne, J.D., Helmer, D., Peters, J. (2005). *The First Steps of Animal Domestication: New Archaeozoological Approaches*. Oxbow, Oxford.

Zeder, M.A. (2012). The domestication of animals. *Journal of Anthropological Research*, 68(2), 161–190. doi: 10.3998/jar.0521004.0068.201.

2

Imprints of Domestication in the Sheep Genome

Charlotte HER and François POMPANON
LECA, Université Grenoble Alpes, France

According to the FAO (Food and Agriculture Organization), there are around 1.4 billion sheep on earth. That is a lot of sheep, but only half the number of nucleotides that make up the genome of each one. If we also add fossil sheep from the past to this, which lived hundreds or thousands of years ago, it is possible to obtain an impressive source of information that can be exploited by geneticists, so as to better understand the origins of these small, domestic ruminants. In this chapter, we will be following the same paths as our ancestors who dispersed sheep across the world. We will move from continent to continent, taking advantage of this journey to illustrate the essential contributions of genetics, in order to understand the domestication process from its origins to the present day.

2.1. The cradle of domestication in the Middle East

2.1.1. *The beginnings of domestication*

Sheep (*Ovis aries*) were domesticated around 10,500 years ago in a region near the Fertile Crescent. Along with goats, they were the first farm animals. However, how do we know this? Firstly, thanks to numerous archaeological findings from the Neolithic period, such as bones, teeth and horns, whose morphology is typical of animals managed by humans. Depending on the age and location of archaeological sites, we can infer the place and time of domestication. However, we need to be sure

Genetics of Domestications,
coordinated by Georges PELLETIER. © ISTE Ltd 2024.

of accurately identifying the species to which the analyzed remains belong. This is a particularly difficult task. Sheep can be confused with goats or their wild counterparts. Furthermore, archaeological remains are often fragmented, especially in the case of animals used for food. Morphometric measurements therefore do not always enable reliable identification. Genetic analysis can therefore help to assign archaeological remains to a species. By comparing DNA sequences, it is possible to distinguish species whose separation has led to the accumulation of mutations that are specific to each. The first genetic study that distinguished between goats and sheep dates back to 1997. Others have since followed, sometimes correcting misidentifications. In 2021, Taylor et al. (2021) extracted DNA from the teeth of four specimens dating from around 5,000–7,000 years before present (BP); these belonged to the genus *Ovis*, according to collagen studies. Several tens to hundreds of thousands of sequence ***polymorphisms*** were identified, and compared with those of modern domestic and wild sheep. The four specimens showed a genetic diversity that was similar to that of modern sheep, but quite different from that of wild sheep. In this case, it was possible to say that these ancient specimens were closer to today's domestic sheep, but it is more difficult to infer their domestic or wild status.

The study of genetic similarities between species also makes it possible to determine which wild species the populations domesticated 10,500 years ago belong to. In the vicinity of documented domestication sites, there are several species of mouflon that could potentially be the ancestor of domestic sheep. These include the Urial (*Ovis vignei*), the Argali (*Ovis ammon*), the Asian mouflon (*Ovis gmelini*)[1] and Urial × Asian mouflon hybrids. While some authors proposed the domestication of sheep from the Argali or Urial, others argued for independent domestication events outside the Middle East, despite the absence of archaeological evidence or present-day wild equivalents in the regions concerned. ***Phylogenies*** can be used as a basis for decision-making. A phylogeny traces the relationships between present-day species or populations on the basis of the resemblance between the individuals that make them up. The closer the groups, the more recent the common ancestor they exclusively share (represented at the divergence node of the branches that represent them). By constructing such trees on the basis of genetic resemblances between sheep and Eurasian wildlife, several studies have shown that sheep are a sister group to Asian mouflon and are well separated from Urials and Argalis (Figure 2.1).

The area of initial domestication can also be inferred from the study of modern genomes. It is possible, for example, to compare the genomes of descendants of the wild ancestor with those of present-day sheep. By identifying and locating the wild

1 The Latin names of Asian mouflon sheep species are in line with the latest IUCN report (Michel and Ghoddousi 2020); however, *Ovis orientalis* is frequently used for this *species*.

individuals that are most genetically similar to the domesticated ones, a potential area of initial domestication can be drawn. According to this approach, the cradle of domestication would extend from North-Western Iran to Eastern Anatolia (Figure 2.2) (Her et al. 2022). This confirms certain archaeological findings and excludes others, such as independent domestication in the Indus Valley. Of course, this assumes that ancient wild populations have not moved too far since the Neolithic period. Today, the Asian mouflon (*Ovis gmelini*), from which sheep originated, occupies a restricted area of north-western Iran, with a few isolated populations further south and in Turkey. Its previously recorded range was not very far from the latter, except that it was larger and less fragmented. It may have spread further south into northern Mesopotamia, perhaps as far as Jordan. On the other hand, it does not seem to have been present on the coasts of the Levant or further north in Turkey, due to the dense forests which are unsuitable to the sheep's needs.

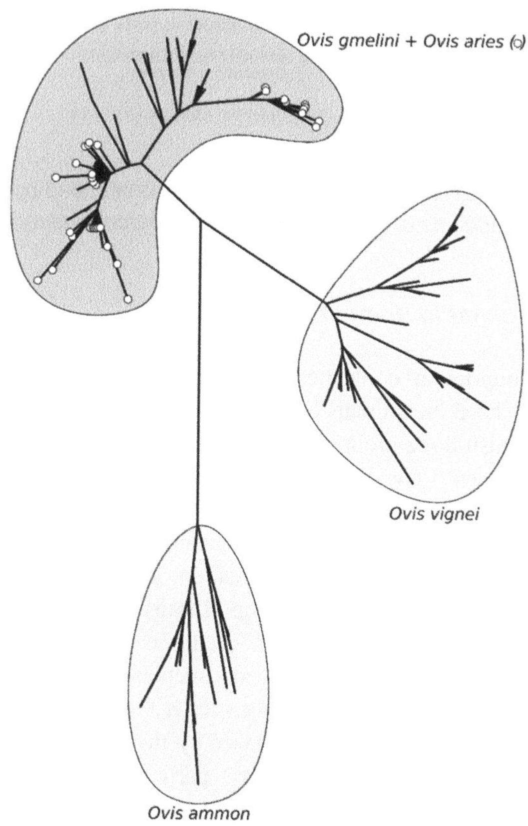

Figure 2.1. *Phylogeny representing the relationships between three species of wild mouflons and domestic sheep (source: illustration by C. Her)*

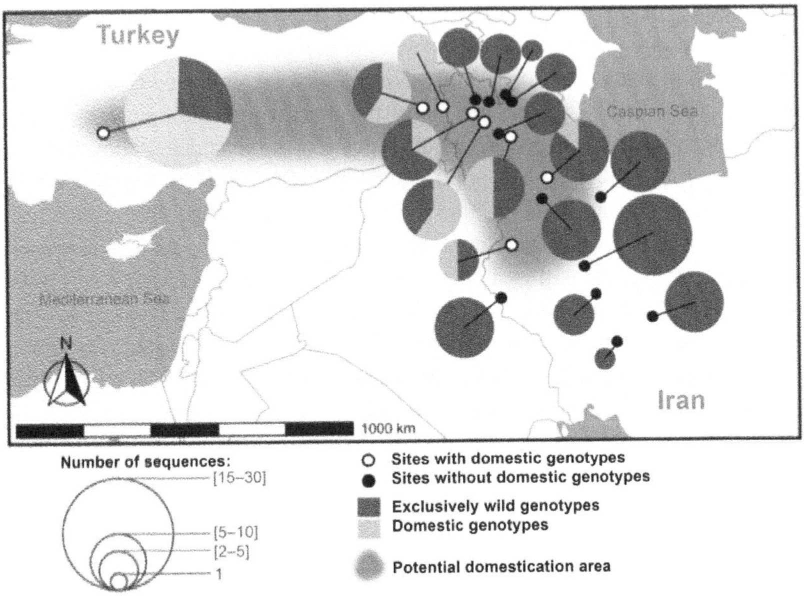

Figure 2.2. *Locating the initial domestication range of sheep by studying the DNA of the descendants of their wild ancestors (source: illustration by C. Her)*

2.1.2. *Genes involved in domestication*

An important imprint of domestication in the animal genome is the selection made by humans to keep the animals best suited to their needs and environment. We can roughly distinguish three groups of traits selected during domestication. Firstly, a set of morphological, physiological and behavioral traits shared by many domesticated mammal species. This group includes features such as floppy ears and distinctive color patterns. Most of these characteristics were probably selected indirectly, during the initial steps of domestication, in connection with the selection of the most docile animals. The second group of traits selected by humans is related to increased economic value. In this case, each species, and even each breed, is purposely selected according to specific needs, such as milk in Friesian milk sheep, or wool in Merinos. The group of characteristics has not been selected by humans directly, but by the selective pressures imposed by the environments into which the animals have been introduced. Conditions, particularly climatic, can be very different from those in the native range of the domesticated species, and thus induce local adaptations.

These selection processes leave an imprint in the genome that can be detected by comparing sheep and their wild counterparts. Indeed, the wild descendants of the ancestral species have evolved independently of the domestic sheep, and we can expect the most differentiated genomic regions between mouflons and sheep to be linked to the domestication process. A second way of finding parts of the genome related to domestication is to identify genomic regions within domesticated populations that are only weakly diversified, given that selection has only favored a few advantageous genetic variants. Several studies have sought to identify domestication genes using these techniques. A first, interesting finding is that a large number of selection signals are found outside the genes (Alberto et al. 2018). These may be variants driven by the selection of a nearby gene variant or involved in regulating the expression of more or less distant genes. Many signals, however, are located in genes involved in multiple biological processes: olfaction, reproduction, immunity, fat metabolism, nervous system, milk or meat production, pigmentation, sexual differentiation, morphology, etc. (Kijas et al. 2012; Alberto et al. 2018). It would take far too long to list all the genes discovered, but we can look at a few.

Let us first take a look at *KITLG*. This gene is involved in several biological processes, such as fleece color, litter size and neural crest cell development. It exhibits selection signals related to domestication in sheep and goats, but which differ in each of these species (Figure 2.3) (Alberto et al. 2018). As *KITLG* is involved in coat color, the signal identified could be explained by directional selection in sheep, for predominantly white wool; whereas, selection for diversification of color patterns would occur in goats, as has already been shown in horses and pigs. This example shows both that there are convergences in the mechanisms of response to selection in closely related species, and that different breeding strategies may have impacted the genome in different ways. Furthermore, this selection signal in the *KITLG* gene in sheep is common to Middle Eastern, European and North African breeds. Therefore, the most parsimonious scenario of selection during the initial stages of domestication, prior to westward diffusion from the center of domestication, is the most likely.

Numerous signals associated with genes involved in zootechnical characteristics were also found. The *CAPN* gene is linked to meat tenderness and growth. *POU1F1* plays a significant role in milk production, notably through the regulation of the growth hormone and prolactin. Wool is also a particularly important trait in sheep. The *KAP* gene family and the *FST* gene are involved in wool quality. The *FST* gene is linked to fiber diameter, while the *KAP* genes regulate hair follicle development. These few examples illustrate the importance of identifying the genetic origin of traits of economic interest for their improvement. Breeding programs based on such studies continue to select animals according to human needs.

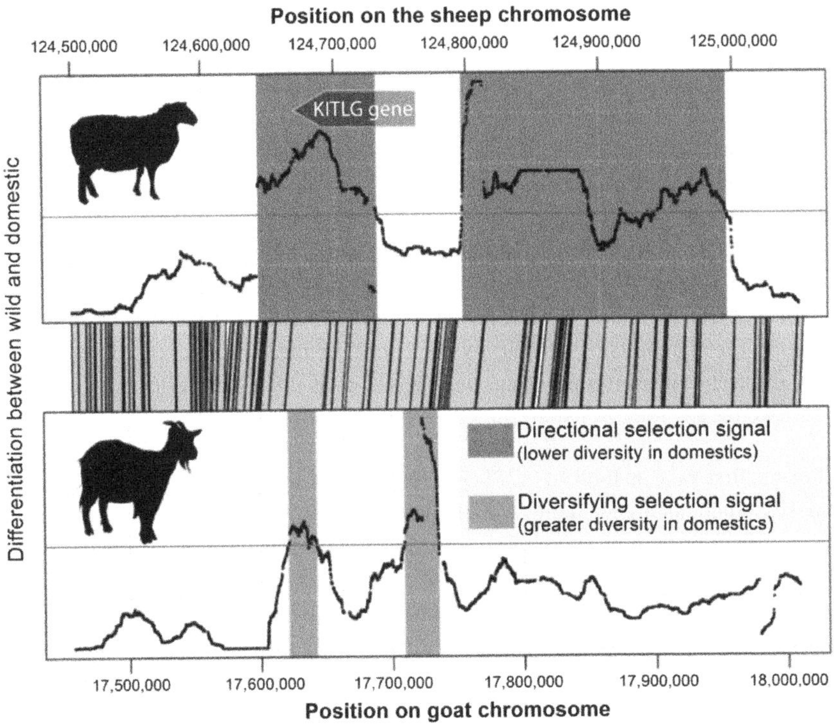

Figure 2.3. *Genomic regions under selection overlapping the KITLG gene (source: illustration by C. Her). For a color version of this figure, see www.iste.co.uk/pelletier/genetics.zip*

COMMENT ON FIGURE 2.3. – *The y-axis represents a measure of genetic differentiation between domestic and wild populations. Genome matches are shown between the two panels and show that the signal on the left includes a strictly homologous region between goat and sheep (source: Alberto et al. 2018).*

While shaping many of the sheep's characteristics, humans have also been the driving force behind its worldwide expansion. This expansion has been complex and illustrates many aspects of domestication's journey.

2.2. Conquering the West

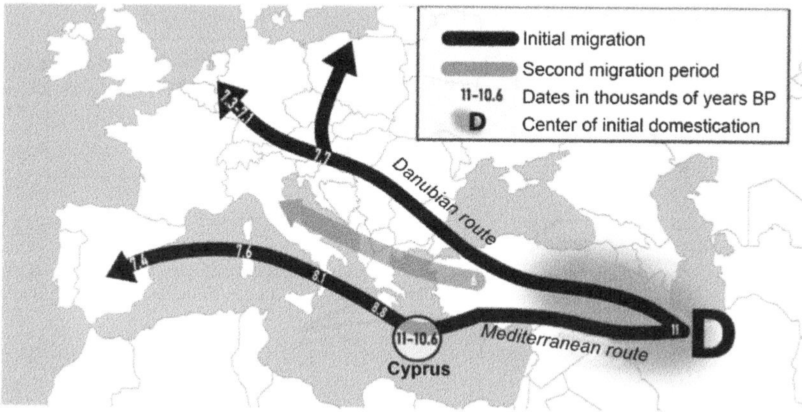

Figure 2.4. *Main sheep diffusion routes in Europe (source: illustration by C. Her). For a color version of this figure, see www.iste.co.uk/pelletier/genetics.zip*

The initial stages of sheep domestication took place in the Middle East, but the animals were soon displaced along the human diffusion routes. Toward Europe, this process followed different routes at different times. The first evidence of the spread of sheep from the center of domestication to Europe was found in Cyprus, along with that of goats, pigs and cows in the second half of the 10th-millennium BP. The rest of Europe was colonized through two routes, the continental Danube route and the Mediterranean route, which crossed the seas via the large islands. The genomes of French sheep are marked by these two influences, as it is in this country that the two routes of diffusion meet. France therefore presents a large proportion of European genetic diversity in a geographically restricted area (Rochus et al. 2018).

Archaeological data and radiocarbon dating show a more rapid arrival in western Europe via the Mediterranean route, around 7,400 years ago (Tapio et al. 2006), supposedly through maritime transport. Moreover, the westward spread of sheep took place in two waves, with the initial movement of domestication dating back 11,000 years and 6,000 years BP (Chessa et al. 2009) (Figure 2.4). This second wave replaced primitive breeds with improved breeds, notably for wool production.

2.2.1. *Primitive sheep versus more productive breeds*

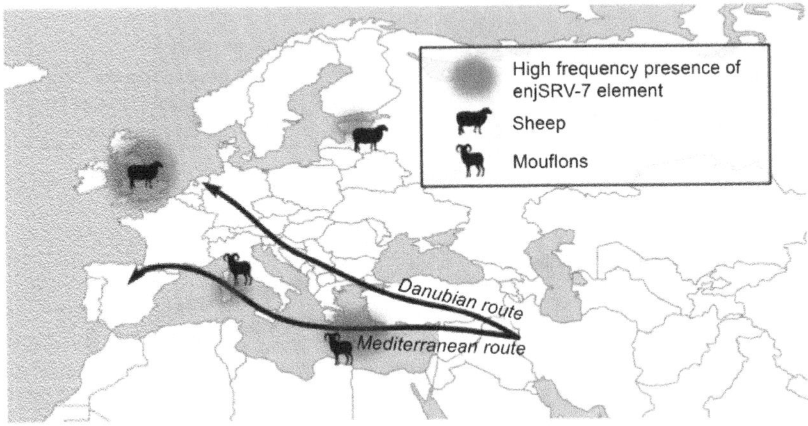

Figure 2.5. *Location of sheep populations with primitive retrotypes, according to Chessa et al. (2009) and Rannamäe et al. (2020) (source: illustration by C. Her). For a color version of this figure, see www.iste.co.uk/pelletier/genetics.zip*

Northern European short-tailed sheep are a group of breeds whose introduction by Neolithic settlers dates back to around 6,000 years ago. Almost all are local breeds, whose current diffusion is close to that of their origin. They are hardy, robust sheep that are highly diversified in terms of size, productivity and performance, and also display archaic characteristics, such as coarse, double-layered wool in various, rather dark colors (which can shed), or the presence of horns, even in females. Examples include the feral Soay, Orkney, Gute and Icelandic sheep. Genetics has confirmed the primitive origin of these breeds in terms of their distribution, thanks to the study of a family of endogenous retroviruses[2] (enJSRV). Each endogenous copy of the enJSRV family results from the stable (transmitted from generation to generation) and irreversible integration of the JSRV retrovirus DNA (free exogenous form) at a different location in the sheep genome, at a different time in the species' history. Therefore, sheep breeds sharing an enJSRV element have a common origin, subsequent to the insertion of this retrovirus. Analysis of the presence or absence of each enJSRV element shows that most European sheep breeds are characterized by the presence of the enJSRV-18 element, with the exception of a few northern European breeds with a primitive morphology, which have the enJSRV-7 element (Chessa et al. 2009) and are therefore distinguished by what we might call

2 A vertebrate-infecting virus whose RNA genome is retrotranscribed into a DNA fragment which, by integrating into the host cell genome, becomes endogenous there.

"retrotypes". These populations are therefore descended from older lineages that spread earlier. The other breeds, which are the majority in Europe, are the result of a second wave of diffusion (Chessa et al. 2009). Replacement by more productive breeds has thus affected most of Europe. Some primitive populations have returned to the wild on islands where predators and humans are scarce, while others have been preserved thanks to their hardiness, which gives them an advantage in harsh environments. However, isolation of these well-adapted breeds is leading to a risk of extinction, as their genetic diversity is deteriorating.

2.2.2. *The mouflons of the Mediterranean islands, relics of the first wave of domestication*

The European mouflon is the result of the ***feralization*** of primitive sheep from the first wave of Neolithic diffusion in Corsica and Sardinia. Chessa et al. (2009) have shown that these populations also carry the enJSRV-7 element, confirming an origin linked to the first wave of Neolithic diffusion. At that time, there were no wild *Ovis* in Europe. The rare incursions of *Ovis* date back to the Middle Pleistocene, several hundred thousand years ago, and involve much larger species whose bones have been found in the mountains of France and Moravia. Today, European mouflons are the only *Ovis* living in the wild in Europe and coexist with herds of domestic sheep. Documents attest to cross-breeding as far back as Roman Antiquity. This results in larger animals that have better reproductive success (Barbato et al. 2017). Such crosses lead to ***introgressions***, transfers of genetic material from one population to another. It has been shown that introgressions from mouflon to sheep can have an adaptive advantage: the genomic regions concerned are involved in immunity and taste recognition (Barbato et al. 2017). Moreover, we find that the genetic contribution of European mouflons, and thus the traces of the first diffusion of the Neolithic era, increases along a gradient from the southeast to the northwest of Europe. This confirms the distribution of primitive retrotypes (Figure 2.5) and coincides with harsher climates, to which introgressions would have enabled adaptation.

Cyprus is another large Mediterranean island occupied by wild sheep. The oldest sheep that remain outside the center of domestication have been found here. These provide evidence of human dispersal, since Cyprus has been isolated from the mainland for over 5 million years and could only have been reached by boat. Cyprus was therefore a stepping stone to Europe during the first wave of marine settlers, dispersing domesticated animals westwards. Present day Cypriot mouflons show very low genetic diversity, due to the long isolation suffered by this population since the Neolithic period, but also due to the intensive hunting carried out in the 1930s, which abruptly reduced its numbers to just 20 individuals, from which the 3,000 that are currently present comes from.

2.2.3. *Southern Europe, diversity and global influence*

Sheep arrived on the Iberian Peninsula from Asia as early as 7,000 BCE, probably via the Mediterranean route. The Iberian breeds were formed as a result of multiple invasions. After the initial wave of Asian immigration, the Celts invaded the region around 2,700 years ago through the Pyrenees. Phoenicians, Greeks, Romans and later Moors followed. The peninsula was thus a commercial and cultural crossroads, integrated into a free-trade zone covering the entire Mediterranean basin, especially in the 11th and 12th centuries. Successive settlers were able to improve local livestock or import some. The most marked changes occurred during the Muslim period, around 1,300 years ago, as evidenced by the significant osteometric differences before and after this period.

Several genetic arguments confirm strong exchanges in this region and suggest a continuous ***gene flow*** via the Mediterranean from the Middle East (Meadows et al. 2005). Indeed, while genetic diversity is expected to decrease through diffusion as we move away from the center of domestication, several Iberian breeds show much higher levels of diversity than other European breeds (Pereira et al. 2006). They also contain genotypes that are usually typical of Asia and the Middle East. The best known of these breeds, the Merino, is mainly bred for its fine, high-quality wool. It has a long history, which is the result of numerous influences from Southern and Eastern Europe, the Near East and North Africa. Its development intensified with the creation of Mesta. This was a breeders' association created in 1273 by Alfonso X of Castile. From this date onward, Merinos were highly protected and their export was strictly forbidden for several centuries. It was not until the 18th century that Merinos were widely exported from Spain. The expansion of Merinos throughout the world is considered to be the factor that has most influenced the development of modern breeds. Kijas et al. (2012) have shown that the majority of European breeds share a considerable genetic component with Merinos. Today, the Merino wool industry is predominantly Australian, while European Merinos are in serious decline, classifying them as an endangered breed. In Australia, intensive genetic selection began in 1870, not just for wool, but also for resistance to disease and parasites (Kijas et al. 2012).

2.3. Africa

Among the 400 breeds listed in Africa, there is a great diversity of sheep phenotypes. While the vast majority of breeds in Europe have wool and fine tails, Africa also has many breeds with fat and/or hairy tails. Thin-tailed sheep are also found in North Africa (from Morocco to Egypt and Sudan) and West Africa (from Senegal to Nigeria). Fat-tailed sheep (Figure 2.6) are found in North Africa, from Algeria to Egypt, and in East Africa, from Eritrea to South Africa. Wool breeds are found in the north of the continent, while hairy breeds are mainly concentrated in sub-Saharan Africa. African sheep have been less studied at the genetic level than Asians and Europeans. However, a number of studies have been carried out to supplement archaeological and historical information.

2.3.1. *Diffusion*

According to the archaeological record, sheep have been present in North Africa for 8,000 years. It is likely that they arrived from the Levant (a region south of the domestication area shown in Figure 2.2) and Sinai, where sheep remains from the same period have been found. Genetic analyses, based on ***mitochondrial DNA***, have revealed two waves of arrivals. The earliest dates back 8,600 years, and the second 7,100 years BP, according to estimates of expansion dates based on genetic data and calibrated by archaeological surveys (Kandoussi et al. 2020). The first would be associated with an early Berber migration in response to the aridification of the Northern Sahara at the end of the African Neolithic Subpluvial (i.e., humid) period. The second came from Iberia. This influence seems to predominate in the genomes of today's Moroccan breeds (Kandoussi et al. 2020). Imprints of this influence can also be found in ancient human genomes of that time, demonstrating the existence of migrations between Europe and Africa, and not just exchanges of livestock. More generally, the study of mitochondrial DNA shows that European and African sheep share a genetic similarity. This was already the case 2,000 years ago, as shown by the genotyping of archaeological specimens (Horsburgh and Rhines 2010). This genetic composition strongly differs from that found in Asia, suggesting a common origin and/or numerous exchanges between European and African populations. In fact, the genome of the Djallonké thin-tailed sheep contains the same retroviral markers as those found in primitive northern European sheep (Chessa et al. 2009) (see section 2.2.1).

Archaeological evidence suggests that sheep reached South Africa later, around 2,000 to 2,500 years ago. They appear to have reached the region by two very different routes (Figure 2.7). The first was via the north, then the west of the continent. This diffusion route is attested by several archaeological discoveries: sheep remains dating back 2,000 years BP in Botswana, and others dating back 2,270 years BP in the Leopard Cave in Namibia. This region may have been a migration corridor during the wetter period, which facilitated the migration of humans and livestock across the Kalahari Desert between 2,500 and 2,000 years ago. The sheep that spread westwards probably had thin tails, as there are no representations of fat-tailed sheep on this side of the continent (Muigai and Hanotte 2013).

Figure 2.6. *Fat-tailed sheep depicted by Job Ludolphus in A New History of Ethiopia (1684)*

The second route runs through eastern Africa, along the mountainous regions of the Rift Valley. There is genetic evidence of this route of diffusion (several genotypes are shared between present-day Kenyan sheep and ancient sheep from Cape Town, South Africa (Resende et al. 2016)), as well as archaeological evidence (sheep remains in the Lake Turkana basin) dating back 4,500–6,500 years BP.

In addition, trade links between Eritrea and the Arabian Peninsula may have provided a secondary entry point for fat-tailed breeds to Africa (Chessa et al. 2009; Muigai and Hanotte 2013). On the other hand, the lack of a shared genotype between Kenyan and Indian sheep would disprove the mass introduction of sheep from southern Asia via the Indian Ocean (Resende et al. 2016).

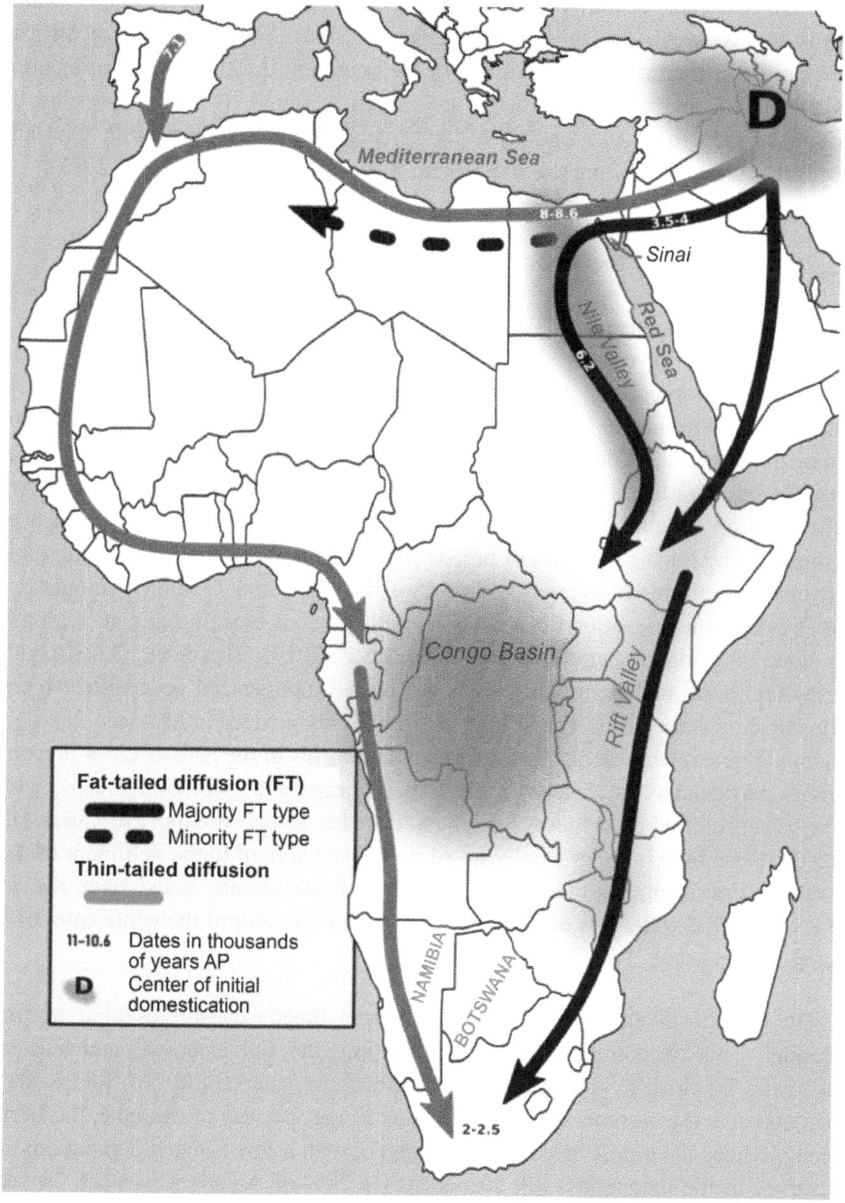

Figure 2.7. *Main sheep diffusion routes in Africa (source: illustration by C. Her). For a color version of this figure, see www.iste.co.uk/pelletier/genetics.zip*

There is a clear disparity between the genetic stocks of East African fat-tailed breeds and West African sheep (Resende et al. 2016). The two routes of diffusion only came together at the southern end of the continent. In Zimbabwe, for example, both fat-tailed and thin-tailed sheep are currently found. It is also possible that fat-tailed sheep from a second wave have replaced or hybridized with thin-tailed breeds that were already present in northern Africa (Muigai and Hanotte 2013).

2.3.2. *North Africa, the impact of the race for productivity*

The Maghreb's sheep herd dates back 7,000 years and contains a wide variety of populations that have been traditionally raised for millennia. This system has enabled the emergence of breeds that are locally adapted to the diversity of the Maghreb's environments. Classification of the main breeds recognized by public authorities is based on morphological characteristics and blood proteins, but there is considerable intra-breed variability. However, as in Europe and the rest of the world, African sheep farming is subject to a race for productivity. This region is a prime example of the homogenization of breeds. There is no overall genetic differentiation between several breeds (Berber, Oulled-Djellal, Rembi and Taâdmit) (Gaouar et al. 2017), and homogenization does not affect all types of breeding and all regions in the same way for the same breed (Belabdi et al. 2019). Therefore, Ouled-Djellal, Timahdite, Sardi and Beni-Guil breeds are more homogenized on private Algerian farms than on state farms. The D'man are more homogenized in Morocco, but not in Algeria. One explanation for this homogenization lies in the lack of cross-breeding. This is particularly the case with the Ouled-Djellal and Beni-Guil breeds, which have been preferred by breeders for several decades, as they are more productive. As a result, these breeds are gradually invading the sheep herd to the detriment of local diversity. For example, the genetic dilution of the Rembi breed over the past 20 years has led to the loss of their bay-tawny color in favor of the white color of the Ouled-Djellal breed.

The genetic diversity present in these local breeds constitutes their adaptive potential. This phenomenon of homogenization, the full extent of which is still unknown, is already raising questions about the adaptability of future sheep populations in the current context of climate change. By way of example, the Berber breed, perhaps the oldest breed in the Maghreb, with a few isolated populations still preserved in the mountains, still accounted for 20% of Algerian livestock 30 years ago. However, the Ouled-Djellal breed has replaced it since 1950. The Barbarine, another ancient breed from the Middle East, is also threatened by the genetic dilution of its heritage and has been in serious decline since the 1990s (Gaouar et al. 2017).

We could follow the spread of sheep westwards to the American continent, but the history of sheep there is recent (from the 15th century onward) and is the result of the introduction of breeds from the Old World. This is why we will move eastwards, and now mention the Asian diffusion routes.

2.4. Asia

The earliest remains of Asian goats have been indirectly dated to around 7,000–5,000 years BP in China, while direct dating of bones and teeth provides dates of 5,700–5,000 years BP (Dodson et al. 2014). The problem with these studies is the uncertainty surrounding the identification of the species from which the anatomically or genetically characterized remains originated. For example, Dodson et al. (2014) identified sheep bones based on short DNA sequences that do not distinguish them from wild *Ovis* (Urial, *Ovis vignei*, and Argali, *Ovis ammon*). The archaeological context of the site and its location are essential in supporting the domestic origin of the animals. An increase in sheep bones can be observed in sites on the central plains of China from 4,500 years ago onwards, and deliberate sheep burials from 5,000 years BP in the Gansu region and from 4,000 years BP when looking further east. From the Zhou dynasty (3,046–2,771 years BP) onward, sheep took on great significance in Chinese culture and were even used in ritual ceremonies. We can therefore say that sheep became widely established in China at the end of the Bronze Age, around 3,000 years ago.

2.4.1. *Genetic signatures of origin, expansion and mixes*

2.4.1.1. *Diffusion*

To reconstruct the stages of sheep diffusion in Asia, Zhao et al. (2017) used an ABC (Approximate Bayesian Computation) analysis, the aim of which is to compare evolutionary or demographic scenarios and estimate the parameters describing their dynamics, such as population size or date of introduction, dates of divergence between populations or their hybridization rate. In practice, genetic data are simulated according to different scenarios, so as to identify the one with the parameters closest to those calculated with real data.

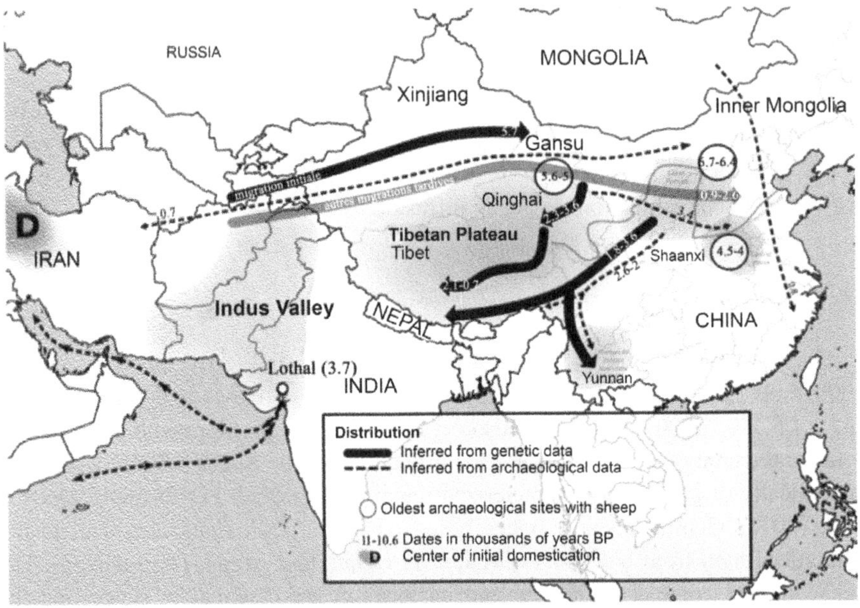

Figure 2.8. *Main sheep diffusion routes in Asia (source: illustration by C. Her). For a color version of this figure, see www.iste.co.uk/pelletier/genetics.zip*

The four scenarios compared in this study have specifically enabled us to study the formation of northern China's sheep, which have diverse genetic compositions. One scenario considered that they derived from an ancestral Eastern population that arrived 4,800 years ago, while the other three argued for a mixture between this ancestral population and sheep from the Middle East. Additionally, the four scenarios differed in terms of the dates of divergence between populations from different regions: northern China, the Tibetan Plateau and the Yunnan-Kweichow Plateau. The best-case scenario involved the individualization of sheep in northern China around 4,000 years ago, predating the separation of sheep from the Tibetan Plateau and Yunnan Kweichow 2,500 years ago. The results also indicate that a later, secondary mix took place between the northern populations and Middle Eastern sheep 1,600 years ago (Figure 2.8). Other elements support this scenario. Populations in the north have a higher genetic diversity than those in the south. This is consistent with earlier divergence and the long history of isolation and founder effects in the Tibetan Plateau and Yunnan Kweichow populations (Lv et al. 2015; Zhao et al. 2017). Furthermore, the presence of certain genotypes that are typical of the Middle East supports the thesis of late hybridization with sheep from this region (Lv et al. 2015). These inferences coincide with historical data, such as the massive population movements during the

Qin wars[3] of unification between 2,000 and 2,400 years BP (Zhao et al. 2017). The arrival of sheep in the Tibetan highlands, dated by genetics to around 3,100 years ago, also coincides with historical archeology information showing the transition to an agro-pastoral system at this period. After colonization, at least 20,000 years ago, human societies settled on the plateau margins between 5,200 and 3,600 years BP, living off millet and livestock, according to archaeological data. Later, between 3,600 and 2,300 year BP, the higher altitudes were colonized and crops changed to more cold-hardy species, such as barley and wheat (Figure 2.8).

2.4.1.2. *Genetic structure linked to different human cultures*

A global trend found in much of the world is low genetic structuring of sheep breeds or populations; in other words, high genetic homogeneity, as we have seen in South Europe and North Africa. However, certain practices completely reverse this trend. On the Tibetan Plateau (encompassing the provinces of Tibet and Qinghai), the nomadic lifestyle is the most common. The Tibetan sheep dominates the plateau, with around 20 million individuals. They provide a wide variety of raw materials, both for primary needs (food, clothing, etc.) and cultural ones (religious totems, ornaments, etc.). Sheep on this plateau have been more isolated, and less cross-bred with exotic breeds, as is the case in northern China, resulting in lower genetic diversity, and strong differentiation from breeds outside the plateau (Zhao et al. 2017; Hu et al. 2019). Similarly, in India, there is very strong genetic structuring due to persistent breeding practices that do not encourage mixing between breeds (Arora et al. 2013). There is still very little movement between Indian states, and each pastoral community follows fixed transhumance routes, helping to maintain breed differentiation.

Conversely, in most of China, there is little structuring between breeds, probably due to extensive exchange and cross-breeding induced by human migration and trade (Cai et al. 2011). Today, mainland China has 67 million sheep and 42 indigenous breeds. However, these breeds only represent a small proportion of the total flock, and 15% of them are in danger of extinction. Their decline dates back some 30 years and has several causes, including the introduction and cross-breeding with numerous exotic breeds in the second half of the 20th century, as part of a national campaign to improve native breeds. A great deal of genetic diversity was lost as a result of unfavorable correlations between traits of economic interest and adaptive characteristics. This phenomenon is similar to the one described for North Africa. In Mongolia, the weak structuring of indigenous breeds can be explained by

---

3 Series of military campaigns launched towards the end of the 3rd century BCE, against the other fighting kingdoms – Han, Zhao, Yan, Wei, Chu and Qi – in the territories forming modern China (https://fr.wikipedia.org/wiki/Guerres_d%27unification_de_Qin).

the traditional method of breeding, involving extensive grazing and seasonal movements favoring the exchange of animals between populations.

2.4.1.3. *Adaptation*

While domestication implies the selection of animals best suited to human needs, the dispersal of animals far beyond their native range has made them face highly diversified environments and climates, causing them to adapt to new and sometimes extreme environments. This is the case, for example, of Tibetan sheep, which live at an average altitude close to 5,000 m. This extreme environment has been the driving force behind major adaptations. Comparison of the genomes of Tibetan sheep with those of lowland sheep shows that several genes are particularly differentiated between these groups. The latter are known to be involved in blood vessel formation, energy production and red blood cell formation (Wei et al. 2016), and would therefore play a role in altitude adaptation. In particular, *EPAS1* plays a significant role in the response to hypoxia and is also associated with altitude adaptation in humans, pigs and dogs. It is likely that Tibetan sheep inherited these adaptations through cross-breeding with a local sheep, the Argali. Hybridization between sheep and mouflons is a common practice among nomadic populations on the plateau (Zhu (2010) in Hu et al. (2019)), and genomic analyses have identified gene variants that are characteristic of mouflons in these sheep. These include genes linked to oxygen metabolism in blood, such as the *HBB* gene (Hu et al. 2019).

2.5. Conclusion

The study of genetic imprints left by the domestication process in sheep genomes provides a better understanding of the events that have shaped breed diversity today. It complements archaeological approaches, so as to trace demographic and migratory events along the routes of diffusion and describe the changes that have shaped these animals over the millennia under the influence of selection by humans and the living environment. Genetics is also a tool for the future development and maintenance of breeds. It is possible to identify genes of agronomic interest and those involved in adaptation to particular local conditions. These approaches are fundamental for conserving the diversity of livestock, preventing the homogenization of breeds and maintaining the genomic resources needed to adapt to the new environmental conditions that organisms will face in the context of climate change.

2.6. References

Alberto, F.J., Boyer, F., Orozco-terWengel, P., Streeter, I., Servin, B., de Villemereuil, P., Benjelloun, B., Librado, P., Biscarini, F., Colli, L. (2018). Convergent genomic signatures of domestication in sheep and goats. *Nat Commun*, 9, 813. doi: 10.1038/s41467-018-03206-y.

Arora, R., Yadav, H.S., Mishra, B.P. (2013). Mitochondrial DNA diversity in Indian sheep. *Livestock Science*, 153(1), 50–55.

Barbato, M., Hailer, F., Orozco-terWengel, P., Kijas, J., Mereu, P., Cabras, P., Mazza, R., Pirastru, M., Bruford, M.W. (2017). Genomic signatures of adaptive introgression from European mouflon into domestic sheep. *Scientific Reports*, 7(1), 7623.

Belabdi, I., Ouhrouch, A., Lafri, M., Gaouar, S.B.S., Ciani, E., Benali, A.R., Ould Ouelhadj, H., Haddioui, A., Pompanon, F., Blanquet, V. et al. (2019). Genetic homogenization of indigenous sheep breeds in Northwest Africa. *Scientific Reports*, 9(1), 7920.

Cai, D., Tang, Z., Yu, H., Han, L., Ren, X., Zhao, X., Zhu, H., Zhou, H. (2011). Early history of Chinese domestic sheep indicated by ancient DNA analysis of Bronze Age individuals. *Journal of Archaeological Science*, 38(4), 896–902.

Chessa, B., Pereira, F., Arnaud, F., Amorim, A., Goyache, F., Mainland, I., Kao, R.R., Pemberton, J.M., Beraldi, D., Steer, M. et al. (2009). Revealing the history of sheep domestication using retrovirus integrations. *Science*, 324(5926), 532–536.

Dodson, J., Dodson, E., Banati, R., Li, X., Atahan, P., Hu, S., Middleton, R.J., Zhou, X., Nan, S. (2014). Oldest directly dated remains of sheep in China. *Scientific Reports*, 4(1), 1–4.

Gaouar, S.B.S., Lafri, M., Djaout, A., El-Bouyahiaoui, R., Bouri, A., Bouchatal, A., Maftah, A., Ciani, E., Silva, A.B.D. (2017). Genome-wide analysis highlights genetic dilution in Algerian sheep. *Heredity*, 118(3), 293–301.

Her, C., Rezaei, H.R., Hughes, S., Naderi, S., Duffraisse, M., Mashkour, M., Naghash, H.R., Bălăşescu, A., Luikart, G., Jordan, S. et al. (2022). Broad maternal geographic origin of domestic sheep in Anatolia and the Zagros. *Animal Genetics*, 53, 452–459. doi: 101111/age.13191.

Horsburgh, K. and Rhines, A. (2010). Genetic characterization of an archaeological sheep assemblage from South Africa's Western Cape. *Journal of Archaeological Science*, 11, 2906–2910.

Hu, X.J., Yang, J., Xie, X.L., Lv, F.H., Cao, Y.H., Li, W.R., Liu, M.J., Wang, Y.T., Li, J.Q., Liu, Y.G. et al. (2019). The genome landscape of Tibetan sheep reveals adaptive introgression from argali and the history of early human settlements on the Qinghai–Tibetan Plateau. *Molecular Biology and Evolution*, 36(2), 283–303.

Kandoussi, A., Boujenane, I., Auger, C., Serranito, B., Germot, A., Piro, M., Maftah, A., Badaoui, B., Petit, D. (2020). The origin of sheep settlement in Western Mediterranean. *Scientific Reports*, 10(1), 10225.

Kijas, J.W., Lenstra, J.A., Hayes, B., Boitard, S., Neto, L.R.P., Cristobal, M.S., Servin, B., McCulloch, R., Whan, V., Gietzen, K. et al. (2012). Genome-wide analysis of the world's sheep breeds reveals high levels of historic mixture and strong recent selection. *PLOS Biology*, 10(2), e1001258.

Lv, F.H., Peng, W.F., Yang, J., Zhao, Y.X., Li, W.R., Liu, M.J., Ma, Y.H., Zhao, Q.J., Yang, G.L., Wang, F. et al. (2015). Mitogenomic meta-analysis identifies two phases of migration in the history of eastern Eurasian sheep. *Molecular Biology and Evolution*, 32(10), 2515–2533.

Meadows, J.R.S., Li, K., Kantanen, J., Tapio, M., Sipos, W., Pardeshi, V., Gupta, V., Calvo, J.H., Whan, V., Norris, B. et al. (2005). Mitochondrial sequence reveals high levels of gene flow between breeds of domestic sheep from Asia and Europe. *Journal of Heredity*, 96(5), 494–501.

Michel, S. and Ghoddousi, A. (2020). IUCN red list of threatened species: *Ovis Gmelini*, Mouflon [Online]. Available at: https://www.iucnredlist.org/en [Accessed 2 September 2020].

Muigai, A.W.T. and Hanotte, O. (2013). The origin of African sheep: Archaeological and genetic perspectives. *African Archaeological Review*, 30(1), 39–50.

Pereira, F., Davis, S.J.M., Pereira, L., McEvoy, B., Bradley, D.G., Amorim, A. (2006). Genetic signatures of a Mediterranean influence in Iberian Peninsula sheep husbandry. *Molecular Biology and Evolution*, 23(7), 1420–1426.

Rannamäe, E., Saarma, U., Ärmpalu-Idvand, A., Teasdale, M.D., Speller, C. (2020). Retroviral analysis reveals the ancient origin of Kihnu native sheep in Estonia: Implications for breed conservation. *Sci. Rep.*, 10, 17340. doi: 10.1038/s41598-020-74415-z.

Resende, A., Gonçalves, J., Muigai, A., Pereira, F. (2016). Mitochondrial DNA variation of domestic sheep (*Ovis aries*) in Kenya. *Animal Genetics*, 47, 377–381.

Rochus, C.M., Tortereau, F., Plisson-Petit, F., Restoux, G., Moreno-Romieux, C., Tosser-Klopp, G., Servin, B. (2018). Revealing the selection history of adaptive loci using genome-wide scans for selection: An example from domestic sheep. *BMC Genomics*, 19(1), 71.

Tapio, M., Marzanov, N., Ozerov, M., Ćinkulov, M., Gonzarenko, G., Kiselyova, T., Murawski, M., Viinalass, H., Kantanen, J. (2006). Sheep mitochondrial DNA variation in European, Caucasian, and Central Asian areas. *Molecular Biology and Evolution*, 23(9), 1776–1783.

Taylor, W.T.T., Pruvost, M., Posth, C., Rendu, R., Krajcarz, M.T., Abdykanova, A., Brancaleoni, G., Spengler, R., Hermes, T., Schiavinato, S. et al. (2021). Evidence for early dispersal of domestic sheep into Central Asia. *Nat Hum Behav*, 5, 1169–1179. doi: 10.1038/s41562-021-01083-y.

Wei, C., Wang, H., Liu, G., Zhao, F., Kijas, J.W., Ma, Y., Lu, J., Zhang, L., Cao, J., Wu, M. et al. (2016). Genome-wide analysis reveals adaptation to high altitudes in Tibetan sheep. *Scientific Reports*, 6, 26770.

Zhao, Y.X., Yang, J., Lv, F.H., Hu, X.J., Xie, X.L., Zhang, M., Li, W.R., Liu, M.J., Wang, Y.T., Li, J.Q. et al. (2017). Genomic reconstruction of the history of native sheep reveals the peopling patterns of nomads and the expansion of early pastoralism in East Asia. *Molecular Biology and Evolution*, 34(9), 2380–2395.

3

Humans and Pigs: Over Ten Thousand Years of Shared Evolution

Laurent FRANTZ[1,2]

[1] *Ludwig Maximilian University of Munich, Germany*

[2] *QMUL, London, United Kingdom*

Pig species share a long history with humans, from their association with the early hominids in Africa to their domestication. Today, they represent an essential source of protein for billions of people and are even used as a biomedical model. In this chapter, we summarize the evolution of this key animal in our history, the distant evolution of its progenitor, the wild boar (*Sus scrofa*), and the recent evolution of commercial breeds that now dominate production in the northern hemisphere.

3.1. The evolution of *Sus scrofa* over the last 2 million years

Domestic pigs (*Sus scrofa domesticus*) belong to the genus *Sus*. To understand the evolution of the domestic pig (*Sus scrofa domesticus*), we first need to understand the evolution of its progenitor, the wild boar (*Sus scrofa*). Indeed, the wild boar is the result of complex genetic mixing over the past 2 million years. Ultimately, it is these genetic variations that have been mixed and selected, over more than 10 thousand years of shared history with humans, to give today's domestic pig.

Genetics of Domestications,
coordinated by Georges PELLETIER.

Sus scrofa, is an extremely versatile species which, at the end of the Pliocene era (5.3–2.6 million years ago), colonized most of continental Eurasia and the Island Southeast Asia, probably replacing most other pig genera, with the exception of *Babyrousa* and sub-Saharan species (e.g. *Phacochoerus africanus*). Following this massive expansion, the genus *Sus* diversified into numerous species, most of which were restricted to Island Southeast Asia, including *Sus barbatus* (Borneo, Sumatra and Malay Peninsula), *Sus verrucosus* (Java and Ba-wean), *Sus cebifrons* and *Sus philippensis* (Philippines), *Sus ahoenobarbus* (Palawan) and *Sus celebensis* (Sulawesi) (Frantz et al. 2016) (Figure 3.1). *Sus* populations found in continental Eurasia are all grouped into a single species: the ancestor of the domestic pig, *Sus scrofa* (Kijas and Anderson 2001).

Recent genome sequencing efforts and fossil discoveries have clarified the evolutionary history of *Sus scrofa* over the last 2 million years, during which this highly successful species colonized the whole of Eurasia and North Africa. During this expansion, which probably began in southeast China (Groenen et al. 2012; Frantz et al. 2013), *Sus scrofa* replaced many other *Sus*, such as *Sus minor* from Europe and China, *Sus strozzi* from Europe, as well as *Sus peii* and *Sus xiaozhu* from China.

Although *Sus scrofa* probably originated in tropical regions of Southeast Asia (Frantz et al. 2013), this species rapidly adapted to a myriad of eco-systems. In <1 million years, *Sus scrofa* has adapted to life at high altitude in the Himalayas (Li et al. 2013), to the northern latitudes of Eurasia (Ai et al. 2015), to the hot, dry climates of the Middle East and North Africa, and to the temperate climates of continental Europe (Groenen et al. 2012). This vast geographical distribution has enabled the species to accumulate a large number of morphological and genetic variations. On the basis of morphological variation, Groves and Grubb (1993) defined four sub-types and 15 sub-species of *Sus scrofa* (Figure 3.1), including:

– the western type, comprising Europe (*S. s. scrofa* and *S. s. meridionalis*), North Africa (*S. s. meridionalis*), the Middle East (*S. s. zybicus*), Central Asia and Siberia (*S. s. attila* and *S. s. nirgripes*);

– the Indian type, including India and Sri Lanka (*S. s. afinis*), Thailand and Myanmar (*S. s. cristatus*) and western India to Iran (*S. s. davidi*);

– the eastern type, including southeast China and Vietnam (*S. s. moupinensis*), eastern Siberia and Mongolia (*S. s. sibiricus* and *S. s. ussuricus*), Japan (*S. s. leucomystax* and *S. s. riukiuanus*) and Taiwan (*S. s. taivanus*);

– the Indonesian type (also known as the banded pig; *S. s. vittatus*) extends over the entire Sunda arc, including Bali, Java and Sumatra, as well as the Malaysian peninsula.

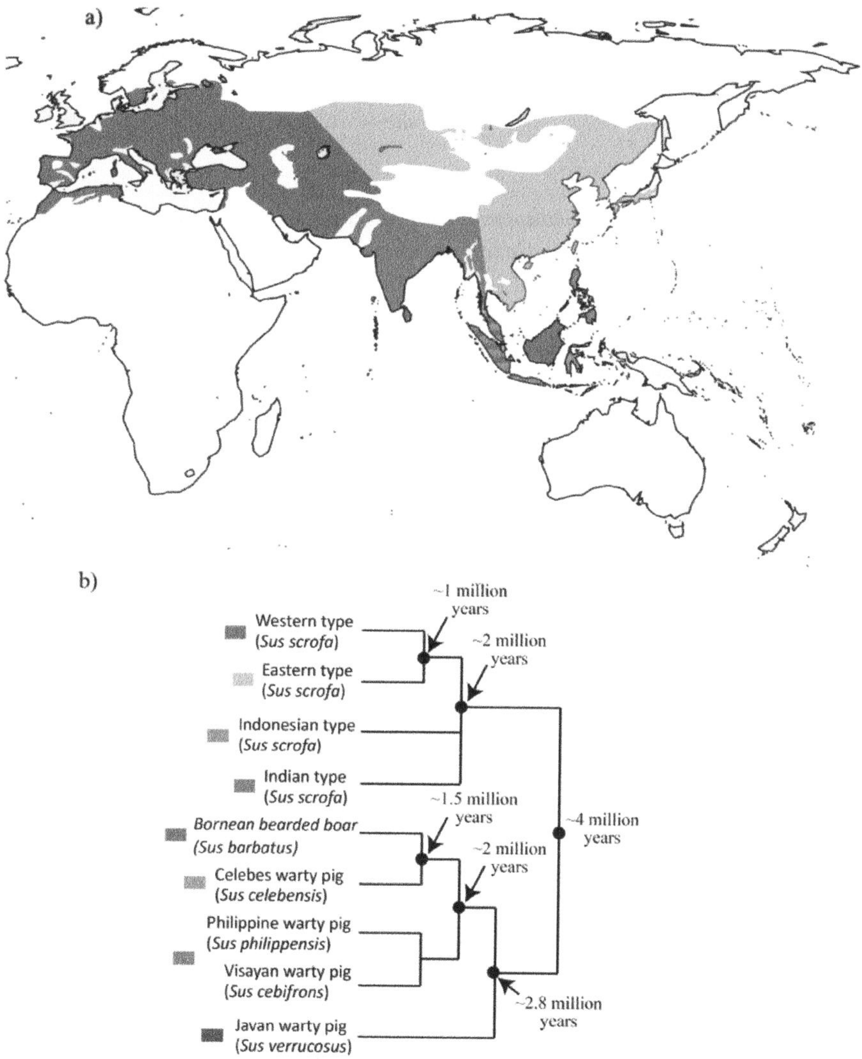

Figure 3.1. *Geographical range and phylogenetic relationships of* Sus *species (source: L. Frantz). For a color version of this figure, see www.iste.co.uk/pelletier/genetics.zip*

COMMENT ON FIGURE 3.1.– *(a) Geographical range of the four types of* S. scrofa: *Western (dark blue), Eastern (orange), Indian (red), Indonesian (green) and the five most common species in insular Southeast Asia:* S. verrucosus *(*Java; red*),* S. barbatus *(Borneo, Sumatra, Malay Peninsula; purple), S. celebensis (Sulawesi; light blue) and* S. philipensis/S. cebifrons *(Philippines; dark orange). (b) Phylogenetic relationship of species/sub-species in Figure 3.1(a). Divergence time represents a rough estimate (Frantz et al. 2013; Liu et al. 2019). The relationship at the base of* S. scrofa *is shown as a polytomy*[1]*, as genome-wide data for Indian types have not yet been published.*

Genetic studies, based on ***mitochondrial DNA*** (Larson et al. 2005), and nuclear studies (Frantz et al. 2013) have made it possible to study the genetic diversity of this species. In particular, recent sequencing efforts have demonstrated the great evolutionary distance between the western, eastern and Indonesian sub-types, which diverged from one another over 1 million years ago (Figure 3.1). However, the geographical origin of *S. scrofa* as a species remains controversial. Genome sequencing studies have been able to show that the western type originated from a migration from northern China around 800,000 years ago, ruling out Europe, North Africa, Central Asia and Siberia as potential geographical origins of the species (Ai et al. 2015). Species of the genus *Sus*, however, are very prone to interspecific ***gene flow*** (cross-breeding between distantly related species), which obscures these genomic analyses.

3.2. The genomics of adaptation in *Sus scrofa*

Genomic studies have also shed light on how *S. scrofa* has adapted to the many environments it has recently colonized. For example, Groenen et al. (2012) suggested that changes in gene regulatory mechanisms (particularly those involving microRNAs) are responsible for the adaptation of *Sus scrofa* during the colonization of Central Asia and Europe. Additional analyses of the variation in gene copy number suggested that genes controlling olfactory perception probably also evolved rapidly during colonization of new habitats (Paudel et al. 2015). This suggests that changes in their ability to smell food in newly colonized environments were an important factor in the success of *S. scrofa* populations (Frantz et al. 2016).

Interestingly, *Sus scrofa* and related species possess over 1,300 olfactory genes, more than in most eutherian mammals sequenced to-date: rats = 1,201; mice = 908; dogs = 872; humans = 457 (Nguyen et al. 2012). Furthermore, recent analyses of pig genomes suggest that variations in the copy number of these genes accumulate more

1 Seemingly simultaneous appearance of several taxa.

rapidly than nucleotide ***polymorphisms*** (**SNPs**) anywhere else in the genome. Furthermore, the number of copies of specific olfactory genes has been shown to be highly selective between species of the genus *Sus* (Paudel et al. 2015). Overall, this has led Frantz et al. (2016) to hypothesize that copy number variation in olfactory receptors has potentially provided suidae with the flexibility to adapt to new environments, contributing to their success since the Eocene era.

Another potential factor in the success of *Sus* species, and *S. scrofa*, in particular, is their tendency to hybridize with other species (Frantz et al. 2016). Interspecific hybridization can, in some cases, be adaptive, as suggested by work in species such as humans and dogs. There are many examples of hybridization that have adaptive potential in *S. scrofa*, particularly in Asia. For example, in Indonesia, on the island of Java, the local species of *S. scrofa* (*S. s. vittatus*) has acquired ~23% of its genome from another species, the endemic *Sus verrucosus*. Interestingly, this ***introgression*** is highly asymmetrical, with only ~5% of the *S. verrucosus* genome derived from *S. scrofa*. This asymmetry is potentially due to the fact that *S. verrucosus* had already been present on the island of Java for millions of years before colonization by *S. scrofa*. *Sus scrofa* therefore potentially acquired new adaptive genetic variations by cross-breeding with the endemic *S. verrucosus* (Frantz et al. 2014).

Similar processes have developed in mainland Asia, such as in China (Ai et al. 2015) and India (Liu et al. 2019). In China, interspecific mixing occurred during the expansion of *S. scrofa* into northern China. Interestingly, the other species involved in this mixing is now extinct. Detecting admixture from extinct species, for which we have no genetic information, is a challenge, as shown by the long-standing controversy over admixture between Neanderthals and modern humans, which was only resolved by the sequencing of the Neanderthal genome. In the case of *S. scrofa*, it is interesting to note that it was the existence of a region of the X chromosome, which has a low rate of recombination at ***meiosis***, leaving an introgression proving hybridization with another species intact for millions of years, that enabled for this interspecific mixing to be revealed. Interspecific hybridization in pigs is particularly important, as the resulting introgressions have potentially provided the genetic material that enabled the expansion of *S. scrofa* and its adaptation to the environmental conditions of northern China (north of the Himalayas) and the rest of Eurasia, and potentially also to adaptation to the human niche (Ai et al. 2015).

Recent sequencing efforts have revealed that interspecific mixing also took place in South Asia between the pygmy hog (*Procula salvanius*) and *S. scrofa*, despite the fact that their morphologies and genomes reflect their divergence of almost 10 million years. However, assessing whether this admixture was adaptive is a challenge that will require the sequencing of many genomes of both Asian *S. scrofa* and pygmy hog (Liu et al. 2019). Such an episode of interspecific admixture, however, could not have been identified without genetic information from the source of the admixture (the pygmy hog). The ability to generate genetic information from extinct species will probably reveal further episodes of interspecific mixing between *S. scrofa* and the myriad of other *Sus* species (e.g. *S. strozzi*) that existed in Eurasia during the Pleistocene era, which was between 2.5 million and 11,700 years BP. Ancient genomes will therefore probably reveal that the rapid expansion of *S. scrofa* was facilitated by a complex pattern of gene flow between species and partial species replacement. It is also possible that these episodes of interspecific gene flow provided the genetic variability necessary for the important production traits of modern pigs.

3.3. The processes of pig domestication

Although dogs were undoubtedly the first domesticated animals, pigs were among the first livestock species to enter a domestic relationship with humans, alongside goats, sheep and cattle. In the Middle East, this process took place around 11,000 years ago, during the establishment of sedentary societies, and our understanding of the history of domestication is now much improved because of advances in empirical data (archaeological and genetic) and the development of new theoretical perspectives.

Historically, researchers, including Darwin, identified morphological and behavioral characteristics that were common to domesticated animals belonging to very different and unrelated species, distinguishing them from their original wild populations. This led to the idea that humans had selected similar characteristics in a wide variety of animals, and that human intentionality had played a significant role in the initial and later stages of domestication.

This interpretation was challenged by Vigne (2011) and Zeder (2012), who have both proposed multi-stage processes: an anthropophilic situation, in other words, adaptation to living in the immediate vicinity of people, followed by commensality, which requires no intentional human action, but takes into account the fact that certain wild populations have been attracted by certain aspects of the human niche, notably food waste. This attraction to people is the start of a process which, in stages characterized by an intensification of the relationship with humans and more direct human control, can result in almost total responsibility for the breeding and survival of a domestic population in captivity.

Another is predation by sedentary human communities who actively hunted the animals. Analyses of bones found at archaeological sites have demonstrated the changing nature of the management strategies used by populations to exploit sheep, goats, pigs and cattle. For example, a study of animal remains at the Sha'ar HaGolan site in Israel showed that cattle and pigs were probably over-hunted before there was any evidence of their domestication (Marom and Bar-Oz 2013). To explain this evolution, it is likely that humans were faced with diminishing animal resources (climate change, and/or increasing human populations) and were forced to adapt their management strategies. This pressure may then have encouraged them to gradually take control of the animals.

It is important to note that the different trajectories followed by domesticated animal species are not mutually exclusive. While it is highly probable that pigs entered a domestication relationship with people when wild boar populations were attracted to the anthropogenic environments and became accustomed to it, this process of commensality does not prevent other wild boar populations from being hunted, thus following the path of predation.

As opposed to the predation and commensal pathway, the directed domestication pathway begins with deliberate intention on the part of people. The animals that fall into this category all entered into a relationship with humans long after the animals that progressed along the paths of commensality or predation. Even for animals such as horses, donkeys and camels, which were presumably more deliberately targeted, their domestication required several generations of genetic adaptation to selective pressures imposed by people. They have, for example, acquired a crucial ability: the ability to reproduce in captivity. Interestingly, the vast majority of domesticated animals in this category were domesticated no more than 1,000 years ago, such as many small pets and aquatic species.

3.4. Using archaeology and genomics to trace the history of pig domestication

Unlike most other domesticated animals (with the possible exception of dogs), pigs have been independently domesticated at least twice: in the Middle East and in East Asia (Giuffra et al. 2000; Kijas and Andersson 2001).

3.4.1. *Archaeology and history of pig domestication*

Archaeological excavations have revealed demographic changes associated with pig domestication at sites in Eastern Anatolia, including Çayönü Tepesi, Hallan

Çemi Tepesi, Hayaz Tepe, Tell Hallula and Gürcütepe. More specifically, the Çayönü Tepesi site has a stratigraphic sequence spanning almost 4,000 years, from 12,000 to 8,300 years BP. An analysis of pig remains, including teeth, from this site has provided evidence that pigs were slaughtered at increasingly younger ages over time. These analyses have also documented morphological changes, including the shortening of the snout and the presence of dental defects resulting from physiological stress. Altogether, these data provided evidence of an intensification of domestic relations. It is important to note that these changes occurred gradually over the whole sequence and that the molar size of ancient pigs did not reach that characteristic of modern domestic pigs until 8,300 years BP (Dobney et al. 2004).

The transition from wild boar to domestic pig may have taken place even earlier at other Neolithic sites, and also in eastern Turkey (Hayaz Tepe, Tell Hallula and Hallan Çemi). At the latter site, a reduction in the length of the third molar has been interpreted as unequivocal proof of the presence of domestic pigs. Although the precise date remains uncertain, zooarchaeological evidence suggests that wild boars were increasingly adapting to the human settlements established in eastern modern Turkey, and that this process led to the appearance of domesticated pigs.

Evidence of an independent domestication of pigs can be found in China. The transition to more intensive farming and sedentary life has been documented from several archeological sites along the Yellow River, around 8,000 years ago, and plant cultivation may have begun 2,000 years earlier. Further south, sedentary hunter-gatherers cultivated rice along the Yangtze River around 11,000–10,000 years BP, and increasing dependence on rice led to more intensive rice cultivation around 8,000 years BP.

There are at least two sub-species of wild boar in China, and this vast territory also possesses a diversity of Neolithic ecological niches and cultural entities scattered across a mosaic environment (Figure 3.1; Groves and Grubb 1993). Previous studies have claimed that domesticated pigs were present as early as 12,500 years BP at the site of Zengpiyan, although this claim has been challenged by a new analysis of zooarchaeological data. The latter analysis also demonstrated a long-term evolution towards domestication at Jiahu (Cucchi et al. 2011), similar to that at Çayönü Tepesi. Although domesticated pigs only represent a small proportion of mammal bones at the earliest sites, they became more common in northern and southern China from 8,000 BCE onwards (Jing and Flad 2002). A new model for the domestication of Chinese pigs suggests that, like in eastern Anatolia, pigs appeared on a scattered regional scale along the Yellow and Yangtze rivers around 8,000 years BP.

3.4.2. *Genomics and the history of pig domestication*

The first genetic approaches made it possible to establish the relationships between variations in domesticated pigs and in Old World wild boars (Larson et al. 2005). These results revealed genetic differences that were closely linked to geographical origin, and confirmed the hypothesis that pigs were domesticated from two distinct populations of wild boar, in Asia and Europe. Subsequent studies have used mitochondrial DNA (Fang and Andersson 2006), nuclear markers such as ***microsatellites*** (Megens et al. 2008) and whole-genome data (Groenen et al. 2012; Bosse et al. 2014; Ai et al. 2015; Frantz et al. 2015). Each of these works has characterized and corroborated the genetic specificity of East Asian and West Eurasian wild boars, as well as the genetic differences observed in domestic pigs derived from these two populations.

3.4.3. *The first study of mitochondrial DNA*

Larson et al. (2005) suggested that pigs were domesticated in more than two different places. The authors suggested this on the basis of a strong phylogeographical signal[2] among many wild boar populations, as well as the observation that domesticated pigs in many regions possessed mitochondrial DNA signatures that matched local wild boar populations. If pigs followed the same path of domestication in different geographical contexts, the long-term demographic and morphological evidence of domestication found in pig remains from archaeological sites in Turkey and China should be apparent in sites in Western Europe, India, peninsular Southeast Asia, and other regions.

However, the archaeological record of these regions reveals a different pattern. Instead of a gradual change in size and demographic profile, domesticated pigs suddenly appeared in Europe and peninsular East Asia at sites associated with migrating farming communities. The archaeological data have been corroborated by ***ancient DNA*** studies of pigs in Anatolia and Europe, which revealed that the first domesticated pigs found in Europe actually possessed mitochondrial ***haplotypes*** associated with Middle Eastern wild boars (Ottoni et al. 2013). In East Asia, ancient mitochondrial DNA signatures from several Chinese sites match those of modern domesticated pigs (Larson et al. 2010). The domesticated pigs associated with Austronesian expansion into the Pacific, however, possess signatures found in modern wild boar from peninsular Southeast Asia (Larson et al. 2007). As in

2 Pertaining to the geographical speciation of a species through the study of its evolution, as a result of migrations based on geography.

Western Europe, archaeological evidence in Southeast Asia does not support a local domestication process either.

In Europe, a few thousand years after the introduction of Middle Eastern pigs, the mitochondrial DNA signature characteristic of Middle Eastern pigs disappeared; it was replaced by haplotypes associated with European wild boars. Recent genomic data from 63 ancient nuclear genomes of Middle Eastern and European pigs revealed that this phenomenon was probably due to gene flow between wild and domestic pigs. The authors showed that the nuclear genomes of European domestic pigs dating from 7,100 to 6,000 years BP had affinities with both Middle Eastern and European wild boars. The majority (around 96%) of these genetic affinities with Middle Eastern wild boars disappeared rapidly after 6,000 years BP (Figure 3.2). This shows that gene flow from European wild boars led to an almost complete disappearance of Middle Eastern ancestry very soon after their introduction. This change was not complete and, for example, ***alleles*** of the *MC1R* gene coding for black coat color, which probably originated in the Middle East, still persist in European pigs.

Therefore, genetic evidence of shared mitochondrial ancestry between local wild boars and domestic pigs from the same region (suggesting widespread and common domestication) contradicts archaeological evidence which, conversely, only suggests two centers of domestication. Studies on the domestication of both animal and plant species have shown that gene flow between native wild populations and transferred domestic populations are very common, and that pigs are not a special case. The result of this mixing is a domestic population that takes on the genetic appearance of a local wild population, although this population has never been domesticated itself. Consequently, early suggestions of multiple centers of domestication were based on an erroneous assumption that genetic affinities between local wild and domestic populations were the result of domestication, rather than secondary gene flow. Furthermore, since exchanges between introduced domesticated populations and local wild populations are so common, the concept of "domestication" should only apply to initial selection processes.

Domestication and subsequent selection by people have generated a host of characteristics that were absent in the original wild animals. Today's European and Asian pig breeds can be distinguished from wild boar by characteristics such as body size, coat color, ear shape and skull shape, as well as reproduction and behavior. However, it is important to distinguish between changes that occurred at the start of domestication, in the strict sense of the term, and more recent changes that have occurred during the development of breeds and intense selection by breeders. The number of genes that have been identified as under selection during specific phases of domestication remains low and does not provide a clear picture of the changes that occurred in the genome during the initial phase of pig domestication.

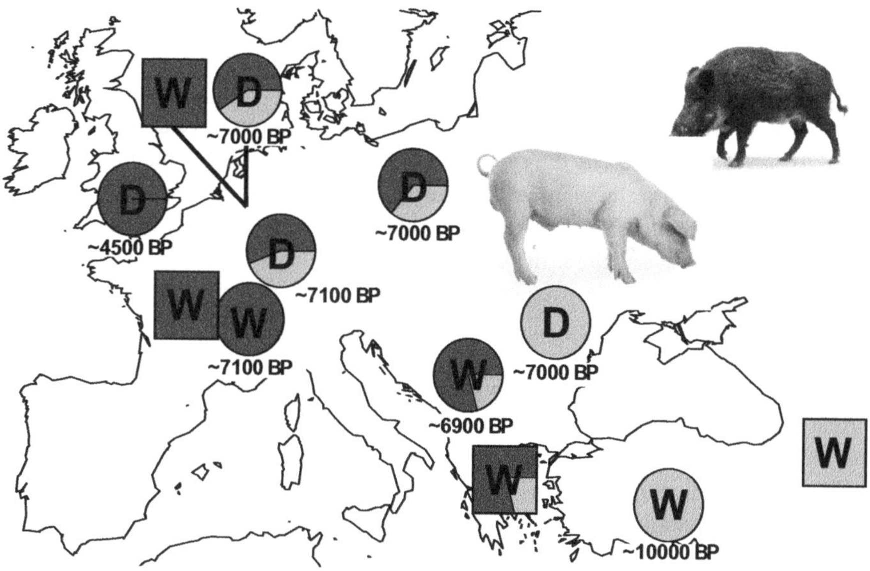

Figure 3.2. *Map showing the ancestry of former domesticated pigs and wild boars (source: L. Frantz). For a color version of this figure, see www.iste.co.uk/pelletier/genetics.zip*

COMMENT ON FIGURE 3.2.– *Circle: ancient genomes, squares: modern genomes. BP = before present. Proportions of ancestry are represented by colors (red: European; yellow: Middle Eastern). Wild samples are 100% yellow or red, while most domesticated animals are hybrids, showing that domesticated pigs transported from the Middle East to Europe crossbred with European wild boars.*

An initial study, based on sequencing data that represented only 2% of the genome, suggested selection on genes that affect coat color, growth, muscle development and harbor potential candidate genes affecting behavior, such as the *PPP1R1B* gene on pig chromosome 12 and the *LRRTM* gene on chromosome 2 (Amaral et al. 2011). A subsequent study, using sequence data covering the whole genomes of several individuals, compared data from domesticated pigs and European wild boar. This approach identified several candidate genes that were selected and that affect body size (*NR6A1*, *PLAG1* and *LCORL*) or body composition (*OSTN*), as well as a mutation in the *NR6A1* gene that is probably responsible for the increased number of vertebrae in domesticated pigs (Rubin et al. 2012). Wilkinson et al. (2013) were able to identify a region on chromosome 1 with evidence of differential selection in domestic breeds compared to wild boar. This selection signal is found in a region containing the *THSB2* and *SMOC2* genes, which have been suggested to affect skull development in dogs and are therefore potential candidates for genes that may have been selected at an early stage during domestication. Behavioral traits were likely under selection during in the early stages of domestication. Genes involved in brain and neuron development are also likely to have been selected during this process. An over-representation of genes linked to neuronal development and neurological regulation has been observed in several studies, although it has not been possible to identify specific *loci* common to different studies. Nevertheless, the complex genetic context of these traits probably involves changes in allele frequency at many loci, rather than a few major domestication loci (Frantz et al. 2015).

Asian breeds are well known for their high prolificacy, fertility and adiposity. Strong ***selective sweep*** signals on genes with a potential effect on fertility (*GPR149* and *JMJD1C*), as well as on genes involved in adiposity traits, have been identified in the Chinese Tongcheng breed. However, these genes were most likely selected well after the first stages of domestication.

The combination of archaeological and genetic evidence has led to two key conclusions. Firstly, although geographically and genetically distinct populations of wild boar exist throughout Eurasia, two independent, long-term processes of pig domestication only occurred in Eastern Anatolia and China. Secondly, following these independent origins, domestic pigs traveled with farmers, were dispersed outside the regions where they had been domesticated and encountered populations of wild boar, with which they hybridized and acquired new mitochondrial and nuclear DNA characteristics. The result of this cross-breeding is that domesticated pigs present in Europe no longer possess the mitochondrial genetic heritage of the Middle Eastern wild boars from which they originated (Ottoni et al. 2013). Furthermore, the first pigs to reach Polynesia carried a mitochondrial signature that is not found in any modern wild boar from the Chinese regions where East Asian pigs were domesticated (Larson et al. 2010).

3.5. The 19th century and the advent of the out stud-book

This process of hybridization between distinct domestic populations continued beyond domestication. For example, the 19th century saw the development of improved breeds (hybrids between Chinese and European pigs). The importation of highly fertile pigs from East Asia during this period had a dramatic impact on the genome of European breeds, such that many European breeds used in commercial settings derive ~30% of their genomic ancestry from East Asian populations (Bosse et al. 2014). More generally, hybridization appears to have been common throughout the natural and artificial evolutionary history of suidae. Therefore, understanding the space-time history of pigs and wild boars, as well as the overall pattern of cross-breeding, will require large amounts of ancient and modern genomic DNA and new bioinformatics approaches.

Cross-breeding between Asian and European domestic stocks, as well as thousands of years of introgression between wild boar and domesticated pigs, have resulted in numerous phenotypes in domestic pigs (Frantz et al. 2015). Intense selection in pigs, however, has only been practiced since the advent of the out stud-book during the industrial revolution of the 18th and 19th centuries, when British breeders crossed Asian pigs with European ones to increase production and meet a growing demand for meat. In China, which has a long history of breeding with countless distinct populations, out stud-books and the concept of breed did not spread at the same time as in Europe, but only more recently. Chinese pigs are often named after their region of origin, most likely due to the great cultural, climatic and topographical variations found in the country.

3.6. Before domestication: human-initiated movements of wild and domesticated pigs

People were also involved in the transfer of domestic pigs and wild boars (as well as other suidae species) around the world. In Africa, humans are thought to have transported Cape warthogs (*Potomachoerus larvatus*) out of the continent to the Comoro Islands and Madagascar (Vercammen et al. 1993). Numerous suidae have also been transported between islands in Southeast Asia. *Babyrousa* may have been deliberately transported from Sulawesi to the islands of Sula and Buru. A suid species native to Sulawesi (*S. celebensis*) was probably transported throughout Island Southeast Asia, as far as the island of Simeulue (northwest of Sumatra), the Moluccas and the Lesser Sunda Islands, including Flores (Blouch 2014), Timor and possibly Bali (Larson et al. 2007). In western Eurasia, wild boar from the Middle East were transported to Cyprus around 11,400 years ago (before they were domesticated), and seafarers are known to have placed wild boar in the Mediterranean islands throughout history.

Domestic pigs were also transported by people across Eurasia and Oceania in prehistoric times, and also more recently. Humans introduced domestic pigs to the Andaman and Nicobar Islands, eastern Indonesia and New Guinea (Larson et al.

2007). European pigs also traveled with the Europeans in the 15th century, who then introduced them to North and South America (Burgos-Paz et al. 2013), New Zealand and Australia. Some of these populations that were introduced have become ***feral*** and invasive, and many now pose a threat to biodiversity, like in Australia. European domestic pigs have also hybridized with local species in Indonesia and the Philippines, where they were introduced (Frantz et al. 2013).

3.7. Conclusion

Today, pork is one of the most widely consumed sources of protein in the world, but only a handful of commercial breeds are used for this production, and many local breeds are threatened with extinction. There is therefore a major risk of losing genetic diversity in this species, which represents a unique biological heritage, forged by over 10,000 years of interaction with people and the environment. The sequencing of genomes, both ancient and modern, should enable the identification of genetic variations that have played a key role in the process of adaptation to local environments, and will serve to prioritize the safeguarding of local breeds. Conserving the full adaptive range that is present in existing genetic resources, and increasing the phenotypic plasticity of commercial breeds, is of major economic and agronomic importance. Advances in the use of genomic techniques are also vital for developing countries, where livestock breeds are not as efficient, often due to environmental constraints. Genomics could support effective breeding programs based on local breeds that are well adapted to environmental conditions, due to thousands of years of co-evolution with people.

3.8. References

Ai, H., Fang, X., Yang, B., Huang, Z., Chen, H., Mao, L., Huang, L. (2015). Adaptation and possible ancient interspecies introgression in pigs identified by whole-genome sequencing. *Nature Genetics*, 47(3), 217–225.

Amaral, A.J., Ferretti, L., Megens, H.J., Crooijmans, R.P., Nie, H., Ramos-Onsins, S.E., Groenen, M.A. (2011). Genome-wide footprints of pig domestication and selection revealed through massive parallel sequencing of pooled DNA. *PLoS One*, 6(4), e14782.

Blouch, R.A. (1995). Conservation and research priorities for threatened suids of South and Southeast Asia. *IBEX JME*, 3, 21–25.

Bosse, M., Megens, H.J., Frantz, L.A., Madsen, O., Larson, G., Paudel, Y., Groenen, M.A. (2014). Genomic analysis reveals selection for Asian genes in European pigs following human-mediated introgression. *Nature Communications*, 5(1), 4392.

Cucchi, T., Hulme-Beaman, A., Yuan, J., Dobney, K. (2011). Early Neolithic pig domestication at Jiahu, Henan Province, China: Clues from molar shape analyses using geometric morphometric approaches. *Journal of Archaeological Science*, 38(1), 11–22.

Dobney, K., Ervynck, A., Albarella, U., Rowley-Conwy, P. (2004). The chronology and frequency of a stress marker (linear enamel hypoplasia) in recent and archaeological populations of Sus scrofa in north-west Europe, and the effects of early domestication. *Journal of Zoology*, 264(2), 197–208.

Fang, M. and Andersson, L. (2006). Mitochondrial diversity in European and Chinese pigs is consistent with population expansions that occurred prior to domestication. *Proceedings of the Royal Society B: Biological Sciences*, 273(1595), 1803–1810.

Frantz, L.A., Schraiber, J.G., Madsen, O., Megens, H.J., Bosse, M., Paudel, Y., Groenen, M.A. (2013). Genome sequencing reveals fine scale diversification and reticulation history during speciation in Sus. *Genome Biology*, 14, 1–12.

Frantz, L.A., Madsen, O., Megens, H.J., Groenen, M.A., Lohse, K. (2014). Testing models of speciation from genome sequences: Divergence and asymmetric admixture in Island South-East Asian Sus species during the Plio-Pleistocene climatic fluctuations. *Molecular Ecology*, 23(22), 5566–5574.

Frantz, L.A., Schraiber, J.G., Madsen, O., Megens, H.J., Cagan, A., Bosse, M., Groenen, M.A. (2015). Evidence of long-term gene flow and selection during domestication from analyses of Eurasian wild and domestic pig genomes. *Nature Genetics*, 47(10), 1141–1148.

Frantz, L., Meijaard, E., Gongora, J., Haile, J., Groenen, M.A., Larson, G. (2016). The evolution of Suidae. *Annual Review of Animal Biosciences*, 4, 61–85.

Giuffra, E.J.M.H., Kijas, J.M.H., Amarger, V., Carlborg, Ö., Jeon, J.T., Andersson, L. (2000). The origin of the domestic pig: Independent domestication and subsequent introgression. *Genetics*, 154(4), 1785–1791.

Groenen, M.A., Archibald, A.L., Uenishi, H., Tuggle, C.K., Takeuchi, Y., Rothschild, M.F., Schook, L.B. (2012). Analyses of pig genomes provide insight into porcine demography and evolution. *Nature*, 491(7424), 393–398.

Groves, C.P. and Grubb, P. (1993). The Eurasian suids: Sus and Babyrousa. In *Pigs, Peccaries, and Hippos*, Oliver, W.L.R. (ed.). IUCN (International Union for Conservation of Nature), Gland.

Jing, Y. and Flad, R.K. (2002). Pig domestication in ancient China. *Antiquity*, 76(293), 724–732.

Kijas, J.M.H. and Andersson, L. (2001). A phylogenetic study of the origin of the domestic pig estimated from the near-complete mtDNA genome. *Journal of Molecular Evolution*, 52, 302–308.

Larson, G., Dobney, K., Albarella, U., Fang, M., Matisoo-Smith, E., Robins, J., Cooper, A. (2005). Worldwide phylogeography of wild boar reveals multiple centers of pig domestication. *Science*, 307(5715), 1618–1621.

Larson, G., Cucchi, T., Fujita, M., Matisoo-Smith, E., Robins, J., Anderson, A., Dobney, K. (2007). Phylogeny and ancient DNA of Sus provides insights into neolithic expansion in Island Southeast Asia and Oceania. *Proceedings of the National Academy of Sciences*, 104(12), 4834–4839.

Larson, G., Liu, R., Zhao, X., Yuan, J., Fuller, D., Barton, L., Li, N. (2010). Patterns of East Asian pig domestication, migration, and turnover revealed by modern and ancient DNA. *Proceedings of the National Academy of Sciences*, 107(17), 7686–7691.

Li, M., Tian, S., Jin, L., Zhou, G., Li, Y., Zhang, Y., Li, R. (2013). Genomic analyses identify distinct patterns of selection in domesticated pigs and Tibetan wild boars. *Nature Genetics*, 45(12), 1431–1438.

Liu, L., Bosse, M., Megens, H.J., Frantz, L.A., Lee, Y.L., Irving-Pease, E.K., Madsen, O. (2019). Genomic analysis on pygmy hog reveals extensive interbreeding during wild boar expansion. *Nature Communications*, 10(1), 1992.

Marom, N. and Bar-Oz, G. (2013). The prey pathway: A regional history of cattle (Bos taurus) and pig (Sus scrofa) domestication in the northern Jordan Valley, Israel. *PLoS One*, 8(2), e55958.

Megens, H.J., Crooijmans, R.P., Cristobal, M.S., Hui, X., Li, N., Groenen, M.A. (2008). Biodiversity of pig breeds from China and Europe estimated from pooled DNA samples: Differences in microsatellite variation between two areas of domestication. *Genetics Selection Evolution*, 40, 1–26.

Nguyen, D.T., Lee, K., Choi, H., Choi, M.K., Le, M.T., Song, N., Park, C. (2012). The complete swine olfactory subgenome: Expansion of the olfactory gene repertoire in the pig genome. *BMC Genomics*, 13(1), 1–12.

Ottoni, C., Girdland Flink, L., Evin, A., Geörg, C., De Cupere, B., Van Neer, W., Larson, G. (2013). Pig domestication and human-mediated dispersal in western Eurasia revealed through ancient DNA and geometric morphometrics. *Molecular Biology and Evolution*, 30(4), 824–832.

Paudel, Y., Madsen, O., Megens, H.J., Frantz, L.A., Bosse, M., Crooijmans, R.P., Groenen, M.A. (2015). Copy number variation in the speciation of pigs: A possible prominent role for olfactory receptors. *BMC Genomics*, 16, 1–14.

Rubin, C.J., Megens, H.J., Barrio, A.M., Maqbool, K., Sayyab, S., Schwochow, D., Andersson, L. (2012). Strong signatures of selection in the domestic pig genome. *Proceedings of the National Academy of Sciences*, 109(48), 19529–19536.

Vercammen, P., Seydack, A.H., Oliver, W.L. (1993). The bush pigs (*Potamochoerus porcus and P. larvatus*). In *Pigs, Peccaries, and Hippos*, Oliver, W.L.R. (ed.). IUCN (International Union for Conservation of Nature), Gland.

Vigne, J.D. (2011). The origins of animal domestication and husbandry: A major change in the history of humanity and the biosphere. *Comptes Rendus Biologies*, 334(3), 171–181.

Wilkinson, S., Lu, Z.H., Megens, H.J., Archibald, A.L., Haley, C., Jackson, I.J., Wiener, P. (2013). Signatures of diversifying selection in European pig breeds. *PLoS Genetics*, 9(4), e1003453.

Zeder, M.A. (2012). The domestication of animals. *Journal of Anthropological Research*, 68(2), 161–190.

4

The Domestication of the Wild Rabbit: Genetic and Genomic Elements

Hervé GARREAU[1] and Cécile CALLOU[2]
[1]GenPhySE, INRAE, Toulouse, France
[2]AASPE, Muséum national d'Histoire naturelle, Paris, France

Rabbit farming provides a wide range of products and services: meat, fur and fiber production, model animals for research, antibody production, pharmaceutical proteins, pets and exhibits. According to the FAO (Food and Agriculture Organization of the United Nations), 893,000 tons of rabbit meat were produced in 2020, with the European Union being the second largest producer, after China, with around 200,000 tons. In the wild, the European rabbit's ability to adapt to a wide variety of environments means that it is generally regarded as a particularly invasive species, overlooking the fact that it is now recognized by the IUCN as a near-threatened species in France and Europe, and even endangered worldwide.[1]

This paradoxical situation and the urge to gain a better understanding of the history of the species, especially that of its domestication, which for a long time remained poorly understood, led to the creation of a research group in the 1990s that brought together both geneticists and archaeozoologists. These initial studies, which combine genetics and ***paleogenetics*** (Monnerot et al. 1994a, 1994b), were based on the existence of abundant populations of wild European rabbits, whose ancestors

1 Available at: https://www.iucnredlist.org.

Genetics of Domestications,
coordinated by Georges PELLETIER. © ISTE Ltd 2024.

were used for this domestication process, and on the discovery of skeletal remains found in well-documented and dated archaeological contexts.

The era of rabbit genomics only began later, with the first complete sequencing of the rabbit genome carried out by the Broad Institute (Boston, USA) in 2009. This fundamental resource paved the way for numerous research projects on the genetics of this small animal. In particular, it has enabled us to discover how domestication may have shaped the animal's genome.

4.1. Phylogenetic context of the species

4.1.1. *Taxonomy*

The placental mammals, or eutherians, are subdivided into three ***clades***: Afrotherians, Laurasiatherians and Euarchontoglires, which, within the Glires clade, include Lagomorphs and rodents. These two orders were separated in 1912 by Gidley, due to anatomical differences, in particular the presence of a second pair of vestigial upper incisors located just behind the first pair of incisors. This separation was confirmed in 1996 by Graur et al. (1996) on the basis of molecular analysis.

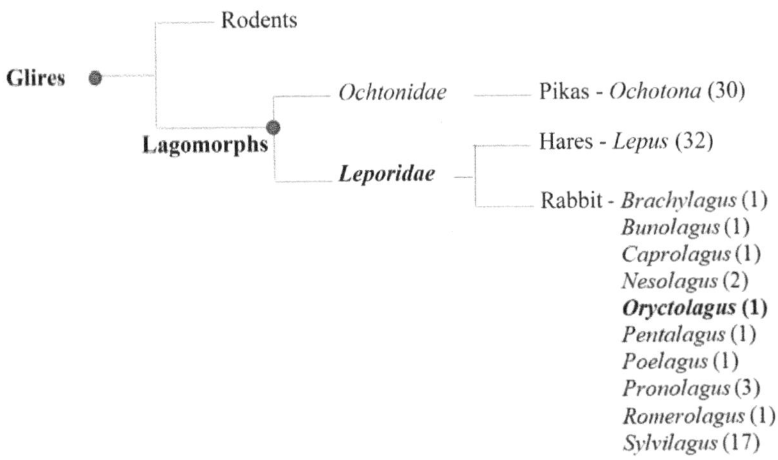

Figure 4.1. *Simplified phylogeny illustrating the families and genera of nonextinct Lagomorphs. The number in brackets corresponds to the number of species per genus (source: from Chapman and Flux (2008, pp. 1–9))*

Lagomorphs are subdivided into two families: pikas belong to the *Ochotonidae* family, which today comprise a single genus, and whose separation from the *Leporidae* is thought to have occurred at least in the Eocene era (50 million years ago); the *Leporidae* family, which includes hares and rabbits, and whose oldest fossil ancestor is thought to have lived in Mongolia 55 million years ago (Asher et al. 2005). The *Leporidae* most probably separated in Asia, then diversified in North America and Asia during the Miocene era, before colonizing South America and Africa (Aulagnier in Savouré-Soubelet et al. 2021).

In his classification of the *Leporidae*, Lyon (1904) noted a great morphological resemblance between the European rabbit *Oryctolagus cuniculus* and the American rabbits of the genus *Sylvilagus* (cotton tails). Nevertheless, according to Corbet (1994), the only justification for separating the two genera lies in the burrowing behavior and extensive, structured social organization of the European rabbit. The taxonomic history of the *Leporidae* and their common ancestors is therefore not fully set in stone.

On the other hand, the existence of two sub-species of the European rabbit *Oryctolagus cuniculus* (Linnaeus 1758) is perfectly accepted and validated: *O. c. algirus* (Loche 1848) and *O. c. cuniculus* (Linnaeus 1758). Representatives of the former subspecies are small (around 1 kg), with very dark coats, and are only geographically distributed in the south-southwest of the Iberian Peninsula. All other geographical areas are occupied by rabbits of the other subspecies, *O. c. cuniculus* (Linnaeus 1758). Larger in size and proportionately heavier (around 2 kg), the individuals belonging to this group have agouti-type coats (gray mixed with ochre, red and black).

Although all authors agree on their existence, the limits of expansion of each of these two lineages remain open to discussion, and do not fully cover the morphological identification of the two sub-species. The populations of the southwestern Iberian Peninsula and the islands to which they were introduced in modern times (Azores, Madeira, etc.) are thus related to lineage A and lineage B, as are the populations of the rest of Spain, Europe and a large part of the world, as well as all domestic rabbits. The discussion mainly focuses on the area of contact and hybridization between the two sub-species, which fluctuates geographically depending on the type of analysis considered, whether mitochondrial, protein or immunological (see, for example, Van der Loo et al. 1991; Monnerot et al. 1994a; Geraldes et al. 2008).

While pikas and hares have the widest natural distribution, rabbits are now the most widespread on the planet.

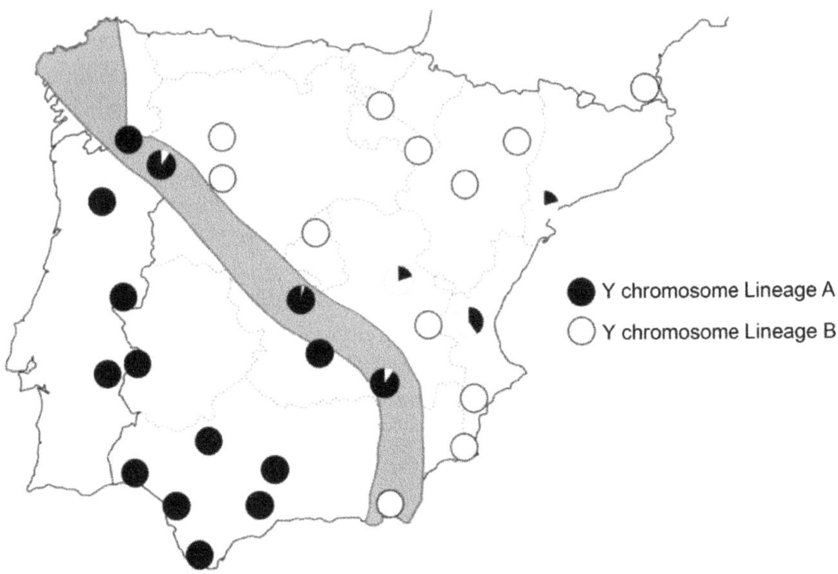

Figure 4.2. *Example of the geographical distribution of A and B lineages in the Iberian Peninsula based on Y chromosome analysis. The presumed hybridization zone is shown in gray (source: illustration by C. Callou, from Geraldes et al. (2008))*

4.1.2. Recent genomics tools

While complete genome sequences for mice, cattle and chickens were obtained between 2001 and 2004, it was not until 2009 that the first rabbit genome was sequenced by the Broad Institute (Boston, USA). As a result of its position in the mammalian ***phylogenetic*** tree and its role as a model species for biomedical research, the rabbit was selected for the Mammalian Genome Project (Lindblad-Toh et al. 2011). The aim of this project was to identify conserved sequences between species in order to improve the annotation of the human genome and produce a comprehensive atlas of functional DNA sequences. Genome annotation, in other words, the location and description of the biological function of genes, was carried out by the *Ensembl* genomics platform, using RNA strand sequencing (RNA-Seq) of rabbit tissues supplied by INRAE and human *ortholog* annotation (by identifying similar genes shared by the two species). The updated genome annotation (*Ensembl release 98*) lists a total of 20,612 protein-coding genes, 8,319 noncoding genes and 51,853 transcripts. The University of Uppsala (Sweden) and the University of Porto (Portugal) have sequenced domestic rabbits from six breeds and French and Spanish wild rabbits, belonging to the two sub-species *O. c. algirus* and *O. c. cuniculus*

(Carneiro et al. 2014). This sequencing identified 51 million ***SNP*** (single-nucleotide polymorphism) markers, as well as 5.6 million insertion/deletion *polymorphisms*. As part of the European project *A Collaborative European Network on Rabbit Genome Biology – RGBNet*, a commercial ***genotyping chip*** (Affymetrix), which supports 200,000 SNP markers, has been developed, enabling high-throughput genotyping of rabbits. It was launched on the market in 2016. In 2019, the European H2020 IMAGE project enabled the development of a multi-species chip (including rabbit, quail, duck, bee and pigeon) with 10,000 SNP markers, mainly dedicated to genetic characterization of populations.

4.2. Origin and spread of the wild rabbit

Among domestic animals, be it mammals or birds, the wild rabbit is a remarkable exception. As the only domesticated Lagomorph, it is also the only domesticated species whose wild origin lies in Western Europe, and more precisely in the Iberian Peninsula (Callou 2003). As with all other domesticated species, several stages have been identified in a long dynamic process, only culminating in domestication a few centuries ago (Irving-Pease et al. 2018).

4.2.1. *Hispanic ancestors*

The first bones belonging to *Oryctolagus cuniculus* were found in southeastern Spain, initially identified at Cullar de Baza (Granada province) and Cueva del Higueron (Malaga province), in the Middle Pleistocene, between 700,000 and 120,000 years BP. Records of bone remains belonging to the species multiplied until the end of the Middle Pleistocene and the beginning of the Upper Pleistocene, mainly in Spain and Portugal, but also in southern France (Callou 2003). Aside from an etymological explanation for the Phoenicians' confusion in identifying them as damans (a species they knew), Spain, or rather the Iberian Peninsula, was indeed the land of rabbits, and remains so to this day. In fact, the greatest genetic diversity was observed in this region.

4.2.2. *Crossing the Pyrenees*

Although the species now occupies all types of habitats, in environments as varied as Australia, the Kerguelen archipelago and over 800 islands, it was originally strictly Mediterranean. Furthermore, the rabbit is particularly sedentary and gregarious, evolving in a fairly restricted home range, limited by a number of environmental constraints, such as altitude and waterways. A priori ill-suited to large-scale colonization, the rabbit has nevertheless managed to leave its original range and cross the Pyrenean barrier, taking advantage of favorable climatic variations and the passage between the province of Girona and the Mediterranean plain of Roussillon. This was only made possible by the displacement of rabbits living in refuge areas during glacial periods, who were able to take advantage of periods of warm, therefore of favorable environmental conditions, during interglacial phases (Callou 2003).

Once over the Pyrenees, the rabbit conquered the whole of the European Mediterranean border as far as Italy until the beginning of the Late Glacial Interstadial, around 18,000 years ago. With the exception of a gradual disappearance from Italy, the natural geographic distribution varied very little on the continent, with the first voluntary transfers of wild rabbits to the islands of the western Mediterranean only beginning in the 14th-century BCE (Balearic Islands; Callou 2003). This initial control in *leporaria* (hare and rabbit pens in Roman antiquity) and movements were undoubtedly the first steps toward future domestication. The fact remains, however, that their purpose was above all hunting: the species is easy to establish and multiplies rapidly without human intervention. There was therefore no real interest in domesticating it, all the more so as some ancient authors insist on the damage it can cause (see Strabo, *Geographica*, III, 2, 6). Is it true that local experiments in captive population control must have been carried out in the Mediterranean region in antiquity, yet is *breeding* the right word here?

4.3. A recent domestication

4.3.1. *From* leporaria *to hutches*

Rabbit distribution underwent a profound upheaval from the Middle Ages onwards. For tens of millennia, rabbits had remained strictly confined to regions with Mediterranean climates and vegetation, but in the space of a few centuries, their geographical limits were broken down, and their place in society rapidly changed. Well established in Western Europe by the 12th century, the rabbit was introduced to the Madeira Archipelago in the 15th century and to South Africa in the 17th century. The first attempts to introduce the rabbit into Australia date back to 1788, but were not successful until 1859. The rabbit was also introduced to the Kerguelen Islands and South America in the 19th century (Callou 2004).

The turning point for the species came in the mid-14th century, when a number of factors contributed to a change in its exploitation, helping to increase its diffusion: the gradual abandonment of the management of warrens (hunting grounds) by lords and ecclesiastics, and the Black Death, which benefited the development of the species through the social upheavals and economic consequences it brought about. With demand for meat and skins on the rise, rabbits were of obvious economic interest, and could no longer be reserved to the upper classes alone. Real population management was put in place, like in England, where an entire population was eliminated for poor "quality", so as to reestablish a new one (Sheail 1971).

From the 13th–14th centuries onward, populations were managed within the warrens, and this was the beginning of animal husbandry. Specific structures were even built to facilitate the "picking" of animals as they left their grounds, as attested by several archaeological discoveries (Le Saint Quinio and Callou 2010). Historical documents also refer to the use of arrangements such as hutches (in the modern sense of rabbit hutches), which enable controlled breeding outside the rabbit warren by isolating certain individuals. In his 16th-century *Historia animalum*, Conrad Gesner reported that:

> There are people in England who earn their living almost entirely by raising rabbits. They allot them a room in their house or a particular dwelling outside the house, and install a crate lined with holes in which the female rabbit gives birth to her young.

The same author also reports having observed variations in coat color.

4.3.2. *The molecular marks of domestication*

Compared with other species that are more widely used in breeding, the number of studies based on the complete rabbit genome sequence is fairly limited. The first studies of genetic comparisons between wild and domesticated rabbits used ***mitochondrial DNA*** sequences. All current breeds of domestic rabbits belong to the mitochondrial B lineage. The mitochondrial types identified in domestic rabbits are sub-groups of the mitochondrial B ***haplotypes*** identified in wild rabbits of the *O. c. cuniculus* sub-species collected in northeastern Spain and southern France (Biju-Duval et al. 1991; Monnerot et al. 1994a; Queney et al. 2002). These initial results confirmed the hypothesis of domestication from populations that have colonized France since the Middle Pleistocene (Monnerot et al. 1994a; Queney et al. 2002; Callou et al. 2003). In a study of 267 domestic rabbits from 24 domestic breeds, Queney et al. (2002) showed that haplotype B1 was the most widespread (249 individuals from all breeds, compared with 18 individuals from seven breeds). Similar mitochondrial profiles have also been observed in domestic populations from China, Egypt and Kenya. This shows that all domestic populations bred worldwide share the same home of domestication and originate from their export from Europe. Subsequent analysis of ***microsatellite*** markers (Alves et al. 2015) and

SNP markers (Carneiro et al. 2011, 2014) confirmed the French origin of domestication: domestic rabbits are genetically closer to French wild than to Spanish wild rabbits.

The reduction in genetic diversity resulting from domestication, which has already been evidenced at the mitochondrial DNA level, is also confirmed by analysis of other markers in the nuclear genome. By comparing the allele frequencies of microsatellite markers in domestic and wild French rabbits, Alves et al. (2015) estimate a 20% reduction in genetic variability. An analysis of SNP markers reveals a reduction in genetic variability twice as high (46%) (Carneiro et al. 2011). Several reasons are put forward by Alves et al. (2015) to explain the contrast between these two values: the statistical methods used in the two studies were different; the representation of the genetic diversity of the domestic populations studied by Carneiro et al. (2011) may have been biased by a smaller number of domestic animals analyzed (n = 25 versus n = 340); lastly, the higher mutation rate of microsatellite markers, compared with SNP markers, may have generated a higher estimate of the genetic diversity of domestic animals. Regardless of the estimated value, this reduction in genetic diversity remains higher than for other species: in dogs, the initial domestication process resulted in a 5% loss of diversity, and the same level of genetic diversity was observed in wild and domestic pigs. The severity of this genetic bottleneck, which is specific to rabbits, is thought to result from a geographically isolated domestication event, based on a small effective number of founders from the only population of wild rabbits of the ***O. c. cuniculus*** sub-species in southwestern France (Carneiro et al. 2011; Alves et al. 2015). This pattern is also consistent with historical and archaeozoological knowledge (Callou 2003).

The most recent studies of genomic data have shown that it is very difficult to accurately date rabbit domestication. The modeling applied by Irving-Pease et al. (2018) aimed to estimate the age of the most recent common ancestor of wild and domestic rabbits. The robustness of the method depends on the rate of mutation, whose reported values in four publications are highly variable (45% between extreme values) and questionable, as inferred from fossil data. Nonetheless, the results suggest that domestic and wild rabbits separated over a period between −12,200 and −17,700 years ago, well before the earliest dates of domestication proposed on the basis of historical or morphological evidence. The authors point out, however, that if the wild populations sequenced to provide the genetic material for the study are not descended from the first domesticated wild animals, the date of divergence of wild and domestic populations may be significantly earlier than the domestication event. In other words, to be more robust, the method assumes that we have genomic data from the wild rabbits from which the first domesticated animals are descended. In conclusion, there is no precise date for rabbit domestication. It should also be remembered that rabbit domestication, like that of other animals, is the result of a gradual and continuous process, linked to the various interactions established between rabbits and humans throughout history.

Data from the complete sequencing of rabbit DNA have been instrumental in revealing the genome changes associated with domestication. In a study published in the *Science* journal, the genomes of six domestic rabbit breeds were compared with those of fourteen wild rabbits collected in southern France and the Iberian Peninsula (Carneiro et al. 2014). Comparison of the allele frequencies of wild and domestic populations identified regions of the genome with lower heterozygosity, this differentiation being the result of the artificial selection carried out in the domestication process. Very few *loci* have been completely fixed in domestic rabbits. With the exception of a few Mendelian traits (discussed below), there are no major domestication genes. However, a very large number of polymorphisms, distributed throughout the genome and affecting gene regulatory regions in particular, have been identified. The results also showed that these polymorphisms affect genes involved in brain and nerve development.

Unlike the domestic rabbit, the wild rabbit has the ability to flee quickly from its various predators (birds of prey, foxes and humans). This study has shown that it is the accumulation of genetic variations with small effects, but on very many genes during domestication, that has progressively inhibited this ability, representing one of the most important changes in the evolutionary history of the rabbit and giving it the notion of domesticity. High-resolution magnetic resonance imaging of wild and domestic rabbit brains supports these findings of selection traces: in domestic rabbits, the volume of the amygdala, implicated in the neurophysiological mechanisms of fear, is smaller, while the volume of the medial prefrontal cortex, which regulates amygdala activity, is higher (Brusini et al. 2018). Domestication would thus have resulted in neurophysiological changes that would have reduced fear and altered flight behavior in wild rabbits.

4.3.3. *Creation of breeds and genes linked to domestication*

The creation of rabbit breeds is a recent phenomenon. One of the oldest is the English silver rabbit, first described in the late Middle Ages. The rabbit fur trade was booming at the time. Certain color variants, resulting from spontaneous mutations, were highly prized by furriers, particularly silver rabbits (a mixture of white hairs on a black background). Towards the end of the Middle Ages, some warrens specialized in breeding silver rabbits (or "Le Riche" rabbits) for the quality and high value of their skins. By the middle of the 19th century, some 10 breeds had been identified. In the second half of the 19th century, amateur breeders in France and England were multiplying the number of ornamental strains, with the number of new breeds growing steadily. It was not until the end of the 19th century that the most complete data on rabbit breeds, their origins and detailed descriptions became available, notably with Darwin's work (1879).

After an initial genetic bottleneck linked to the initial domestication process (see section 4.3.2), the creation of the breeds now generates a further reduction in diversity which is, on average, among the breeds studied, estimated at 20% (Alves et al. 2015). This high level of reduction, similar to that estimated for the initial domestication process, is a result of the high intensity of selection applied at the time of breed creation and reveals a low number of effective ancestors for each of the breeds. However, this value is close to those estimated for the creation of breeds of other domestic species, such as cattle and hens. Analysis of the complete sequences (Carneiro et al. 2014) also reveals that the level of diversity in wild rabbits is very high, making this species one of the most polymorphic of all sequenced mammals. The apparent loss of diversity linked to domestication and the creation of breeds must therefore be put into perspective by considering the richness of the initial genetic reservoir of wild populations. The phylogenetic tree inferred from the microsatellite marker genotyping of rabbits of 19 breeds clearly shows differentiation between breeds (Alves et al. 2015): animals of one breed are genetically closer to animals of the same breed than to animals of other breeds. Although the process of domestication is recent, the significant differences in allele frequencies between breeds can be explained by the species' very short generation range and the strong selection pressure associated with the creation of breeds.

The main consequences of domestication are the appearance and conservation of new hereditary morphological, physiological or behavioral traits of interest to man. These traits, which differentiate domesticated animals from their wild ancestors, are grouped together under the term "***domestication syndrome***". Sequence data have enabled us to identify the causal mutations of a number of genes involved in biological processes of interest for the development of domestic rabbits. As it would take too long to draw up a complete list, we suggest mentioning a few genes here that contribute more particularly to the uniqueness of domestic rabbit breeds, as a result of artificial selection.

The main source of phenotypic diversity in the domestic rabbit, and one of the first traits selected by man (as shown by the example of the English silver rabbit), is coat color. Among the few mutations associated with color, those identified in the *TYR* gene have highly visible effects that immediately distinguish wild rabbits from several domestic breeds. This gene encodes the tyrosinase enzyme that catalyzes the formation of melanin, responsible for hair pigmentation. ***Alleles*** at the Albino locus of the *TYR* gene determine the amount of melanin produced, or even its total absence in albino rabbits. Early in the 20th century, the study of Albino and Brown genetic types (Castle 1924) led to the identification of the first genetic linkage group in this species (Group I and Group LG 1), which was subsequently localized on chromosome 1. Sequencing of the *TYR* gene has also made it possible to identify several mutations corresponding to different alleles at the *Albino* locus (Aigner et al. 2000): full development of the black and yellow pigments in the coat and the black color of the eye, specific to wild rabbits, are determined by the dominant *C* allele. The other alleles are all recessive to *C* and are characterized by variable pigment production. In Chinchilla rabbits carrying the c^{ch} allele, the middle part of the coat is discolored. The effect of the Himalayan (c^{H}) allele is temperature dependent: melanin production is reduced, and pigments only color the extremities of the animal (ears, nose, paw tips), whose temperature is lower than that of the body. The Himalayan and Californian breeds carry the c^{H} allele. The effect of the *Albino* (*c*) allele, visible in New Zealand animals, is reflected in the total absence of melanin, a pink pupil and a white iris. The breeds or lines most commonly used for meat production are the New Zealand and Californian breeds. These two breeds, created in the United States at the beginning of the 20th century, were selected for their zootechnical qualities (prolificacy, maternal aptitudes and growth) and their suitability for rearing on wire mesh. The white color of their coats, complete for the former breed and partial for the latter breed, is appreciated by breeders, as it facilitates observation of the animals and detection of signs of disease, such as diarrhea.

The Rex rabbit is a breed that was officially recognized in 1929. The coat of this rabbit is characterized by a reduction in the number and length of the coarser hairs, giving a coat essentially composed of down, with shorter, more compact and softer hairs. As a result of its pleasant feel, the value of its skin is much higher than that of other rabbit breeds. This recessive trait is the result of a spontaneous mutation, whose effects were first described in 1919. This mutation was identified by Diribarne et al. (2011). It involves a deletion of a nucleotide base in exon 9 (1362delA) of the *LIPH* gene, resulting in a truncated *LIPH* protein. The *LIPH* protein is an enzyme that catalyzes the production of 2-acyl lysophosphatidic acid (*LPA*), which is a lipid mediator with diverse biological properties that include platelet aggregation, smooth muscle contraction and stimulation of cell proliferation. Low levels of *LIPH* expression affect the migration and proliferation of keratinocytes, the precursors of hair formation, leading to a reduction in the number and length of hairs.

Present in 31% and 22% of French households, respectively, cats and dogs are the most widespread pets. However, among the other mammals sold in pet shops, in the "rodents" category, rabbits account for 66% of market share, far ahead of hamsters and gerbils (14%), guinea pigs (9%), and mice and rats (8%). The smallest rabbit breed is the Netherland dwarf. It is one of a number of dwarf rabbit breeds specially developed for this new type of market. Dwarf rabbits generally weigh less than a kilo and have an altered craniofacial morphology. This particular phenotype is governed by the effect of a single gene (Carneiro et al. 2017). The causal mutation of this trait is a deletion of the promoter region and the first three exons of the *HMGA2* gene, leading to the inactivation of this gene. *HMGA2* conditions the expression of the *IGF2BP2* (insulin-like growth factor 2) gene, which is itself involved in the regulation of genes encoding mitochondrial proteins. The same genetic mechanism affecting growth and craniofacial development has been demonstrated in pygmy mice. Genetic diversity analyses of the *LCORL-NCAPG* locus surprisingly suggest that the small size of dwarf breeds has been enhanced by crosses with wild rabbits.

4.4. Back to the wild: an invasive species

Of European origin, the rabbit's presence on all five continents is linked to human activity. Its ability to adapt to a wide variety of environments makes it an outstanding invasive species, having successfully colonized the natural environments of many countries following their introduction by man. The rabbit's expansion into Australia, which is particularly devastating for endemic flora and fauna, and is the cause of agricultural losses estimated at 200 million dollars a year, is the largest animal invasion in recent history. Ninety imports of rabbits have been recorded since their first arrival in this country in 1788, but it is only after 1860 that their progress grew exponentially, with distances of over 100 km covered per year. History also tells us that 24 rabbits from England were released on December 25, 1859, near Victoria, by a settler named Thomas Austin, originally from Barwon Park in South West England.

The genome study of wild Australian rabbits identified three genetically distinct population types: the majority of animals (n = 57), collected from multiple locations across the country, constituted the dominant group. The other two groups were made up of a smaller number of animals from more isolated geographical areas (Alves et al. 2022): one group came from the Sydney region (n = 4) and the other from Cattai National Park (n = 1). Comparison of the genomes of these three groups with those of wild French, English (South West and East of England) and domestic rabbits confirmed their different genetic origins: the genetic profile of Sydney rabbits is very close to that of domestic rabbits. This result is consistent with the first introduction of domestic rabbits to Sydney in 1788. However, the Cattai rabbit and rabbits from the dominant group were genetically closer to English rabbits. In spite of multiple introductions, Australia's invasion would therefore only be the result of the arrival of the 24 or so animals reported in 1859. These animals, captured from wild populations on Thomas Austin's property in South West England, were probably more adaptable than the domestic rabbits that arrived before 1859.

New Zealand and Tasmania followed a similar pattern of expansion. The genetic profile of Tasmanian rabbits is much closer to that of domesticated rabbits and Sydney rabbits, but also reveals ancestral origins from the dominant group of Australian rabbits (Alves et al. 2022). According to historical data, rabbit expansion in Tasmania occurred after the arrival of Thomas Austin's animals in Victoria. Tasmanian rabbits would therefore have been imported from Australia after 1860, and their gene pool is a mixture of dominant Australian and domestic populations. The genetic structure of New Zealand rabbits also reveals a close relationship with domestic rabbits and Australian dominant group rabbits, but in varying proportions depending on the samples taken. Analyses also reveal common ancestry with the wild populations of eastern England, which are genetically distinct from the southwestern populations from which the rabbits reported by Thomas Austin were derived. All of these elements suggest that the New Zealand rabbit population was formed by multiple imports from Australia, but also directly from the East of England.

Pedro Andrade (personal communication) compared complete genomic sequences of domestic, Spanish and French wild rabbits with those of ***feral*** populations introduced in Australia, New Zealand, Argentina, Sweden and the Falkland Islands, in order to study the genetic footprints of ***feralization***. The respective proportions of domestic and wild ancestral contributions to the gene pool of feral populations were estimated. It shows that the domestic ancestral contribution is predominant in most introduced populations, with values ranging from 47.5% (Swedish rabbits) to 80.8% (Argentine rabbits). The allele frequencies of nine causal mutations associated with coat coloration, hair length and texture, and fat tissue coloration in rabbits were studied. Three causal mutations were totally absent in the feral populations (rex, angora and albino). The other mutations were detected, but with low frequency. It seems fairly obvious that the mutations that give the rabbit a highly visible color (albino) and a very particular coat structure (rex, angora) make the animals more vulnerable in a natural environment and are therefore totally nonexistent in feral populations, while the other, less penalizing mutations are segregated with significant frequencies. To investigate a possible mechanism of counterselection of domestic traits, a comparison of the allele frequencies of all feral populations with those of domestic animals has enabled us to identify the most differentiated genomic regions: the most significant differences are found for

25 independent genomic regions, including 86 protein-coding genes. Most of these genes are associated with behavioral traits, as well as the nervous system development and function.

These traits were previously identified by Carneiro et al. (2014) in the first genomic study conducted to identify the genetic markers of domestication (see section 4.3.2). Genome analysis reveals that it is the traits that condition the rabbit's survival in a natural environment (escape behavior and camouflage to escape predators) that differentiate the domestic rabbit from the wild rabbit the most, or the rabbit that has become wild again after a domestication stage.

4.5. Conclusion

The information provided by genomics, combined with archaeological and historical data, has enabled us to better describe the major stages in the spread and domestication of rabbits, from the appearance of the genus in the Miocene era to the contemporary period. After remaining confined to the Mediterranean region for several millennia, the wild rabbit invaded Western Europe, then many parts of the world in the space of a few centuries, with humans being solely responsible for this widespread distribution. As with the chicken, the taming of the rabbit by humans has modified the genetic architecture of brain development, as well as the first major genes associated with its domestication, and the creation of breeds are those that govern the color of its coat. A final feature of the rabbit is that it has no scientific name to distinguish the wild from the domestic form (Gentry et al. 2004).

4.6. References

Aigner, B., Besenfelder, U., Müller, M., Brem, G. (2000). Tyrosinase gene variants in different rabbit strains. *Mammalian Genome*, 11, 700–702.

Alves, J.M., Carneiro, M., Afonso, S., Lapes, S., Garreau, H., Boucher, S., Allain, D., Queney, G., Esteves, P.J., Bolet, G. et al. (2015). Levels and patterns of genetic diversity and population structure in domestic rabbits. *PLOS ONE*, 10, e0144687.

Alves, J.M., Carneiro, M., Day, J.P., Welch, J.J., Duckworth, J.A., Cox, T.E., Letnic, M., Strive, T., Ferrand, N., Jiggins, F.M. (2022). A single introduction of wild rabbits triggered the biological invasion of Australia. *Proceedings of the National Academy of Sciences*, 119(35), e2122734119.

Asher, R.J., Meng, J., McKenna, M.C., White, J.R., Dashzeveg, D., Rougier, G., Novacek, M.J. (2005). Stem Lagomorpha and the antiquity of Glires. *Science*, 307, 1091–1094.

Biju-Duval, C., Ennafaa, H., Dennebouy, N., Monnerot, M., Mignotte, F. (1991). Mitochondrial DNA evolution in lagomorphs: Origin of systematic hetmoplasmy and organization of diversity in European rabbits. *Journal of Molecular Evolution*, 33, 92–102.

Brusini, I., Carneiro, M., Wang, C., Rubin, C.J., Ring, H., Afonso, S., Blanco-Aguiar, J.A., Ferrand, N., Rafati, N., Villafuerte, R. et al. (2018). Changes in brain architecture are consistent with altered fear processing in domestic rabbits. *Proceedings of the National Academy of Sciences of the United States of America*, 115, 7380–7385.

Callou, C. (2003). *De la garenne au clapier. Étude archéozoologique du Lapin en Europe occidentale.* Mémoires du Muséum National d'Histoire Naturelle, Paris.

Callou, C. (2004). L'apport de l'archéozoologie à l'étude de la domestication du lapin. *Ethnozootechnie*, 75, 9–15.

Carneiro, M., Afonso, S., Geraldes, A., Garreau, H., Bolet, G., Boucher, S., Tircazes, A., Queney, G., Nachman, M.W., Ferrand, N. (2011). The genetic structure of domestic rabbits. *Molecular Biology and Evolution*, 28, 1801–1816.

Carneiro, M., Rubin, C.J., Di Palma, F., Albert, F.W., Alföldi, J., Martinez Barrio, A., Pielberg, G., Rafati, N., Sayyab, S., Turner-Maier, J. et al. (2014). Rabbit genome analysis reveals a polygenic basis for phenotypic change during domestication. *Science*, 345, 1074–1079.

Carneiro, M., Hu, D., Archer, J., Feng, C., Afonso, S., Chen, C., Blanco-Aguia, J.A., Garreau, H., Boucher, S., Ferreira, P.G. et al. (2017). Dwarfism and altered craniofacial development in rabbits is caused by a 12.1 kb deletion at the HMGA2 locus. *Genetics*, 205(2), 955–965.

Castle, W.E. (1924). On the occurrence in rabbits of linkage in inheritance between albinism and brown pigmentation. *Proceedings of the National Academy of Sciences of the USA*, 10, 486–488.

Chapman, J.A. and Flux, J.E.E. (2008). Introduction to the Lagomorpha. In *Lagomorph Biology: Evolution, Ecology and Conservation*, Alves, P.C., Ferrand, N., Hackländer, K. (eds). Springer, Berlin.

Corbet, G.B. (1994). Taxonomy and origins. In *The European Rabbit. The History and Biology of a Successful Colonizer*, Thompson, H.V. and King, C.M. (eds). Oxford University Press, Oxford.

Darwin, C. (1879). *De la variation des animaux et des plantes à l'état domestique.* Reinwald et Cie, Paris.

Diribarne, M., Mata, X., Chantry-Darmon, C., Vaiman, A., Auvinet, G., Bouet, S., Deretz, S., Cribiu, E., De Rochambeau, H., Allain, D., Guerin, G. (2011). A deletion in exon 9 of the LIPH gene is responsible for the rex hair coat phenotype in rabbits (*Oryctolagus cuniculus*). *PLoS One*, 6(4), e19281.

Gentry, A., Clutton-Brock, J., Groves, C. (2004). The naming of wild animal species and their domestic derivatives. *Journal of Archaeological Science*, 31, 645–651.

Geraldes, A., Carneiro, M., Delibes-Mateo, M., Villafuerte, R., Nachman, M.W., Ferrand, N. (2008). Reduced introgression of the Y chromosome between subspecies of the European rabbit (*Oryctolagus cuniculus*) in the Iberian Peninsula. *Molecular Ecology*, 17, 4489–4499.

Graur, D., Duret, L., Gouy, M. (1996). Phylogenetic position of the order Lagomorpha (rabbits, hares and allies). *Nature*, 379, 333–335.

Irving-Pease, E.K., Frantz, L.A., Sykes, N., Callou, C., Larson, G. (2018). Rabbits and the specious origins of domestication. *Trends in Ecology and Evolution*, 33, 149–152.

Le Saint Quinio, T. and Callou, C. (2010). Une "garenne à lapins" aux Grands Sablons à Crêches-sur-Saône (71). *Groupement archéologique du Mâconnais info*, 1, 1–13.

Lindblad-Toh, K., Garber, M., Zuk, O., Lin, M.F., Parker, B.J., Washietl, S., Kheradpour, P., Ernst, J., Jordan, G., Mauceli, E. et al. (2011). A high-resolution map of human evolutionary constraint using 29 mammals. *Nature*, 478, 476–482.

Lyon, M.W. (1904). Classification of the hares and their allies. *Smithsonian Miscellaneous Collections*, 45, 321–447.

Monnerot, M., Vigne, J.D., Biju-Duval, C., Casane, O., Callou, C., Hardy, C., Mougel, F., Soriguer, R., Dennebouy, N., Mounolou, J.C. (1994a). Rabbit and man: Genetic and historic approach. *Genetics Selection Evolution*, 26(1), 167–182.

Monnerot, M., Casañe, D., Hardy, C., Vigne, J.D. (1994b). Evolution of *Oryctolagus*: Relationships within the *Lagomorpha* system and access to population history by ancient DNA. *Polish Ecological Studies*, 20(3/4), 543–551.

Queney, G., Vachot, A.M., Brun, J.M., Dennebouy, N., Mulsant, P., Monnerot, M. (2002). Different levels of human intervention in domestic rabbits: Affects on genetic diversity. *Journal of Heredity*, 93, 205–209.

Savouré-Soubelet, A., Arthur, C., Aulagnier, S., Body, G., Callou, C., Haffner, P., Marchandeau, S., Moutou, F., Saint-Andrieux, C. (2021). *Atlas des mammifères de France. Ongulés et Lagomorphes*. Muséum national d'Histoire naturelle, Paris.

Sheail, J. (1971). *Rabbits and their History*. Newton Abbot, London.

Van Der Loo, W., Boussès, P., Arthur, C.P., Chapuis, J.L. (1996). Compensatory aspects of allele diversity at immunoglobulin loci: Gene correlations in rabbit populations devoid of light chain diversity. *Genetics*, 144, 1181–1194.

5

Domesticated Poultry: A History Illuminated by Genomics

Michèle TIXIER-BOICHARD[1], Xavier ROGNON[1] and Bertrand BED'HOM[2]

[1]*GABI, INRAE, AgroParisTech, Université Paris-Saclay, Jouy-en-Josas, France*

[2]*ISYEB, Muséum national d'Histoire naturelle, Paris, France*

5.1. Introduction

Domesticated species are differentiated from the wild species they originate from through the gradual acquisition of specific characteristics throughout the domestication process by the human species, the latter controlling the environment and reproduction of domesticated populations through the selection of breeding individuals. These species can be classified schematically according to the services they provide to humans: companionship or ornamentation, work or leisure, or production (mainly food).

The list of domesticated species can vary according to sources and criteria, notably the age of domestication. A decree issued by the French Ministry of Ecology[1] defines a domestic species as "a species whose representatives belong to selected animal populations or are descended from parents belonging to selected animal populations". This decree sets out the list of species or breeds/varieties officially considered to be domestic in France. However, this list seems excessively detailed for domestic birds. Indeed, while the wild ancestor species is still present, the list specifies each domestic variety, including its color. Furthermore, the decree does not distinguish between species that have been domesticated for at least

---

1 Available at: https://www.legifrance.gouv.fr/jorf/id/JORFTEXT000000789087 (2006).

Genetics of Domestications,
coordinated by Georges PELLETIER. © ISTE Ltd 2024.

1,000 years, and those that have been domesticated for only a few dozen to a few hundred years. The latter are essentially pet, aviary or ornamental species, especially passerine and psittacine birds. The species considered in this chapter are those with a long history of domestication, and which are currently mainly bred for food production (meat, eggs): chicken (*Gallus gallus*), guinea fowl (*Numida meleagris*), turkey (*Meleagris gallopavo*), quail (*Coturnix japonica*), two species of duck – the domestic duck (*Anas platyrhynchos*) and the domestic Muscovy duck (*Cairina moschata*), two species of geese – the greylag goose (*Anser anser*) and the Chinese goose (*Anser cygnoides*), and the domestic pigeon (*Columba livia*). Aside from model species such as the zebra finch (*Taeniopygia guttata*), these are also bird species for which we have powerful genomic tools, enabling us to study domestication, population structuring or the effects of selection. Following a presentation of bird ***phylogeny***, we will discuss domestication scenarios for the various species, before turning to the genetic mechanisms involved.

5.2. Domestic birds and their phylogenetic context

Recent genomic studies have resolved many of the *phylogenetic* uncertainties between different orders of birds. Birds are the only dinosaurs to have survived the mass extinction of the late Cretaceous (66 million years ago) and have subsequently undergone a relatively rapid diversification of different evolutionary lineages. This peculiarity has made it difficult to identify phylogenetic relationships between certain orders with limited molecular data. Access to whole-genome sequencing has improved phylogenetic reconstructions, with nodes between major bird lineages now much better resolved and supported (Kuhl et al. 2021). A simplified bird *phylogenomic* tree, in which groups that include species domesticated over a long period of time are indicated by arrows, is shown in Figure 5.1.

Although the evolutionary history of today's birds has seen some 30 orders emerge with very different specific diversities (only one species today, the hoatzin, for the order Opisthocomiformes, and over 6,000 for the order Passeriformes), species with a long history of domestication are concentrated in three orders (Figure 5.1): columbiformes (domestic pigeon), galliformes (chicken, guinea fowl, turkey and quail) and anseriformes (ducks and geese). Galliformes and anseriformes are also sister groups, forming the monophyletic *Galloanserae* lineage, which includes species from which 110 billion of the world's domesticated birds are raised for consumption every year.[2] The main domestic species are thus derived from just four families (Figure 5.1): *Columbidae*, *Phasianidae*, *Numididae* and *Anatidae*, the latter three having strong phylogenetic similarities.

2 2019 FAOSTAT data (https://www.fao.org/faostat/en/#data/QCL).

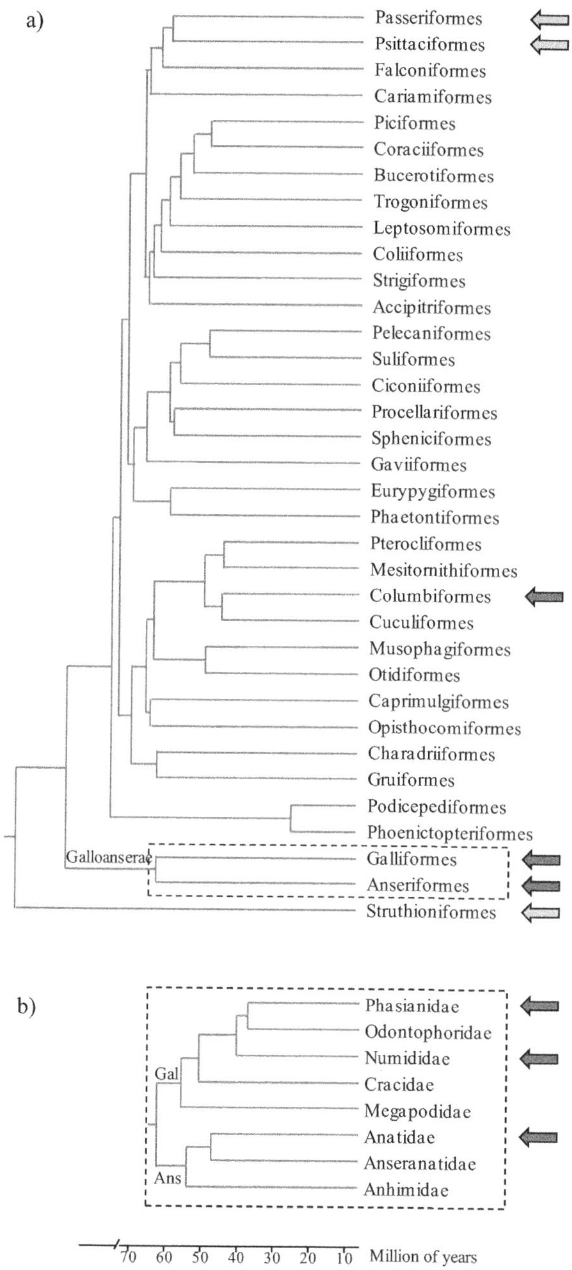

Figure 5.1. *Phylogeny of birds (adapted from Kuhl et al. (2021)). The length of the branches indicates the divergence time in millions of years*

COMMENT ON FIGURE 5.1.– *Dark arrows indicate groups of species domesticated at least 1,000 years ago, and light arrows indicate groups of species that have recently been domesticated. (a) Phylogeny of bird orders. The Galloanserae group is surrounded by dotted lines; (b) phylogeny of Galloanserae families, galliformes (Gal) and anseriformes (Ans).*

Bovidae, in mammals, and galliformes, in birds, are prevalent among domesticated production animals, and certain biological characteristics of many species in these groups would favor domestication, notably social structure (hierarchical gregarious groups dominated by one male), precocial (nesting) young, short flight distance and a herbivorous or omnivorous diet. The wild species domesticated were hunted prey or commensal species to humans. These "pre-adaptation" traits to domestication are partially phylogenetically constrained, which helps to explain the limited diversity in the evolutionary origins of domesticated species.

The phylogenetic context of domesticated species is also interesting to consider in terms of interspecific hybridization. The ability to produce viable interspecific hybrids is generally greater in birds than in mammals. Among birds, the frequency of interspecific hybridization is particularly high in anseriformes and galliformes, with 41.6% and 21.5% of species in these orders, respectively. Some species even have the ability to hybridize (naturally or in captivity) with a large number of species: 39 for the mallard (*Anas platyrhynchos*) and 14 for the common pheasant (*Phasianus colchicus*). This natural capacity for hybridization in galliformes and anseriformes may lead to the persistence of ***gene flow*** after speciation, through occasional hybridization between closely related species, insofar as hybrids can be fertile, even at a low rate. The possible persistence of gene flow does not make it easy to reconstruct the initial phylogenetic relationships of wild species that are close to domestic species.

The development of genome analysis tools has made it possible to revisit the study of domestication and gain a better understanding of ancestral wild species, as well as the dynamics involved. The assembly of partial genomic sequences leads to the construction of complete reference genomes, which are available for several species (Table 5.1) in dedicated databases (such as the NCBI database, National Center for Biotechnology Information), where they can be viewed using specific search tools. In this table, we can see that these genomes have two levels of assembly quality: a whole chromosome assembly ("chromosome") or a discontinuous assembly, at the scaffold scale ("segment" of chromosome) alone. The latest assembly technologies also make it possible to phase the genome of the sequenced individual in the form of two parental ***haploid*** assemblies. One or more reference genomes (in other words, an assembled and annotated genome) have been produced for each species. They can be updated, with improved assembly or annotation from one version to the next. These reference genomes are indispensable resources for studies in phylogenomics, population genomics, ***selection signatures*** and genetic mapping.

Species	Version	Year	NCBI identifier	Assembly	Status
Gallus gallus	6	2018	GCF_000002315	Chromosome	Reference
	1	2017	GCA_002798355	Chromosome	–
	1	2021	GCF_016699485	Chromosome, haploid	Reference
	1	2021	GCA_016700215	Chromosome, haploid	–
Coturnix japonica	2	2020	GCF_001577835	Chromosome	Reference
	2	2016	GCA_000511605	Segment	–
Numida meleagris	1	2017	GCF_002078875	Chromosome	Reference
	1	2020	GCA_014595935	Segment	–
Meleagris gallopavo	3	2019	GCF_000146605	Chromosome	Reference
	1	2021	GCA_905368555	Chromosome	–
Anas platyrhynchos	1	2020	GCF_015476345	Chromosome	Reference
	1	2013	GCF_000355885	Segment	–
	1	2018	GCF_003850225	Chromosome	–
	1	2017	GCA_002224895	Segment	–
	1	2019	GCA_008746955	Chromosome	–
	1	2021	GCA_017639285	Segment	–
	1	2021	GCA_017639305	Segment	–
	1	2020	GCA_900411745	Chromosome	–
	1	2017	GCA_002743455	Chromosome	–
Cairina moschata	1	2021	GCA_018104995	Chromosome	Reference
	1	2019	GCA_009194515	Chromosome	–
Anser cygnoides	1	2015	GCF_000971095	Segment	Reference
	1	2017	GCA_002166845	Segment	–
	1	2020	GCA_013030995	Segment	–
Columba livia	2	2017	GCF_000337935	Segment	Reference
	1	2016	GCA_001887795	Chromosome	Reference

Table 5.1. *Status of genome data for major domesticated bird species. The prefixes GCA and GCF in the identifiers indicate GenBank and RefSeq assemblies, respectively (source: NCBI (National Center for Biotechnology Information[3]))*

3 Available at: https://www.ncbi.nlm.nih.gov/.

Genomic tools have thus enabled us to explore long-standing phylogenetic questions in greater detail, such as relationships within the genus *Gallus*, which comprises four species found in Southeast Asia: Red junglefowl (*G. gallus*), Grey junglefowl (*G. sonneratii*), Sri Lanka junglefowl (*Gallus lafayettii*) and Green junglefowl (*G. varius*). *Gallus gallus* has the widest range, stretching from southeast India and southern China to Indonesia, while the other three species occupy smaller areas, with *G. varius* on the island of Java and surrounding islands, *G. lafayettii* in Sri Lanka and *G. sonneratii* in a region extending from central to southern India. The phylogenetic relationships between the four species of the genus *Gallus* (*G. gallus*, *G. sonneratii*, *G. lafayettii* and *G. varius*) are still the subject of conflicting studies, depending on the populations used, the genetic data and the methods employed. Even the most recent studies (Lawal et al. 2020; Mariadassou et al. 2021), which use whole-genome data, do not agree on a single topology. The main uncertainty lies in the most basal branch of the tree, constituted by *G. gallus* or *G. varius*. According to the studies, whose sampling and methods differ, the most basal species of the tree is either *G. gallus* (Mariadassou et al. 2021) or *G. varius* (Lawal et al. 2020). The analysis carried out by Mariadassou et al. (2021) separates the two trees with a significantly greater likelihood for the tree with *G. gallus* in the basal position (Figure 5.2). This relative controversy could be resolved by a joint analysis of the data from these two studies.

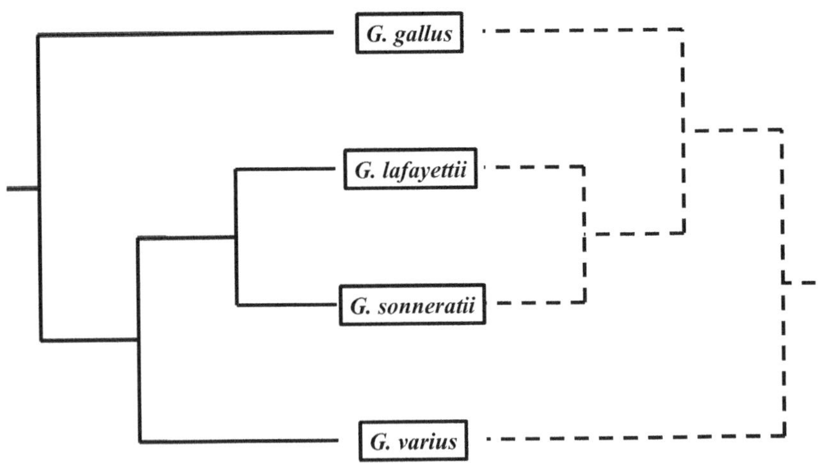

Figure 5.2. *Most probable topologies of phylogenetic relationships between present-day species of the genus Gallus, with G. gallus on the left and G. varius on the right as the most basal branch (dotted lines)*

The *G. gallus* species is itself subdivided into five sub-species, whose differentiation is still under debate. Mariadassou et al. (2021) observe that the *murghi* and *bankiva* sub-species are each distinct from the other two sub-species analyzed in this study, *gallus* and *spadiceus*, which do not clearly appear to be separated, while Wang et al. (2020) show a significant differentiation between each of the five sub-species. These studies identify interspecific gene flow through hybridization during the evolution of the species, which contributes to the uncertainties in reconstructions of the evolutionary history of this genus. The subject of gene flow after domestication is discussed in section 5.3.

In anseriformes, hybridization capacity may also have played a role in the domestication of geese and domestic ducks, for which genomic contributions from more than one species are assumed for the constitution of domestic populations. In addition, high interspecific hybridization capacity in anseriformes is exploited for the production of hybrid ducks (mulards) between relatively distant species belonging to two different genera, the domestic duck (genus *Anas*) and the Muscovy duck (genus *Cairina*), with these sterile hybrids being used for the production of foie gras.

At a higher taxonomic level, recent phylogenomic studies have confirmed the evolutionary relationships within the Galliformes order, already well established by studies based on specific mitochondrial data. The turkey thus belongs to a *clade* of the *Phasianidae* family, with the chicken and quail belonging to another clade of the same family, and the guinea fowl to the more basal *Numididae* family.

There is currently no phylogenomic study of the order anseriformes. Only the goose group (in the strict sense) has been studied, thanks to genome-wide genetic data, which resolved the topological uncertainties of previous studies that had insufficient information of genetic markers due to the context of rapid diversification of geese, and probably hybridization between differentiated species.

5.3. Domestication scenarios

In this section, dedicated to genomics' contribution to the understanding of domestication scenarios, we will first look at the chicken, the species with the most archaeological and genetic information available; and, in particular, the significant development of studies based on the analysis of whole genomes. We will then look at a number of other species domesticated for farming, such as guinea fowl, turkey, quail, the domestic duck and the domestic pigeon. There are currently no genomic studies on the domestication of geese and Muscovy ducks.

5.3.1. *The chicken*

5.3.1.1. *The parent species*

Among domesticated bird species, the chicken (*Phasianidae* family) is the one for which there is the most data, both archaeological and genetic, for the study of domestication scenarios. Of the four species described in the previous section, *G. gallus* is the closest to domesticated chickens (for a review, see Tixier-Boichard and Weigend (2020)). This is true from both a morphological and a reproductive point of view; crosses between domestic forms of chickens and *G. gallus* produce viable, fertile offspring without any difficulty; whereas, those involving other species have low success rates. These data, together with analyses based on ***mitochondrial DNA polymorphism*** or combining mitochondrial and nuclear DNA, have led us to consider *G. gallus* as the wild ancestor of the chicken. However, the contribution of other species in the genus *Gallus* has also been studied using molecular approaches. For example, ***introgressions*** from *G. sonneratii* into domesticated forms of *G. gallus* have been demonstrated, and their signature can be seen in the "yellow skin" phenotype present in certain breeds and lines of domesticated chickens. This phenotype is controlled by the *BCO2* gene. A study of the phylogenetic relationships between the different variants in *Gallus* species for the region carrying this gene (23.8 kb) shows the existence of two groups; one grouping the sequences present in *G. sonneratii* and *G. lafayettii* ("yellow skin"), the other carrying the sequences present in *G. gallus* ("white skin"). The sequences found in the domestic species are associated with one or other of the clades, in line with the phenotype observed. A study based on genomic data (Lawal et al. 2020) shows significant gene flow between *G. sonneratii* and domestic populations (12% for the autosomal nuclear genome, and 6% for the Z chromosome). It should be noted, however, that the *G. sonneratii* sampling in this study only includes individuals born and bred in captivity. Mariadassou et al. (2021) have shown that most of the world's captive *G. sonneratii* populations are the result of relatively recent hybridization (around 150 years), with around 10% of their genome derived from *G. gallus*. Therefore, to estimate the rate of introgression of *G. sonneratii* in domestic chicken populations, it is necessary either to use genome information from wild *G. sonneratii* individuals or, if they come from parks, to search for and remove regions introgressed by *G. gallus* (Mariadassou et al. 2021). Lawal et al. (2020) also highlight the possible contribution of *G. lafayettii* and *G. varius* to the domestic chicken genome. Conversely, it is interesting to note that gene flow between domestic and wild populations of *G. gallus*, and between different species of the genus *Gallus*, has also been reported in various studies, based on mitochondrial DNA polymorphism, ***microsatellite*** and genomic data (Tixier-Boichard and Weigend 2020; Mariadassou et al. 2021).

5.3.1.2. *Locations and period of domestication(s)*

Approaches based on mitochondrial polymorphism have been used to study domestication in chickens for almost 30 years. Initially, these were mitochondrial DNA sequencing studies of wild or domesticated individuals, either focusing on a fragment of the mitochondrial genome for a large number of individuals, or on the entire mitochondrial genome for a smaller number of individuals (Tixier-Boichard and Weigend 2020). These studies concluded in favor of independent domestications in different locations, including Southwest China, Southeast Asia and South Asia.

Archaeological data have suggested at least two possible regions of domestication. One is in the Indus Valley, with remains dating from 4,000 to 4,500 years before present (BP), but outside the species' range, indicating that domestication may have been earlier. The other, in North China, dates back to 8,000 years BP. However, the identification of most of the bone remains from Neolithic and Early Bronze Age sites in North China has been called into question: only a very small proportion of these represent the remains of domestic chickens or red junglefowl, while the majority correspond to other *Phasianidae* species. These data are consistent with the environmental data (climate, flora, etc.) of the region during this period, not compatible with the living conditions of *G. gallus*, which is a tropical species. Based on the study of mitochondrial DNA from bone remains, Barton et al. (2020) show that the birds exploited at the Dadiwan site in northern China at this time were the common pheasants (*Phasianus colchicus*). In a recent study, Peters et al. (2022) re-analyzed data from over 600 archaeological sites using a conservative approach that removes a lot of ambiguous data. As a result, only three sites in Thailand were identified as the oldest with a confirmed presence of domestic chickens, the oldest being the Ban Non Wat site, dating from 3,650 to 3,250 years BP. These findings are consistent with original domestication in this area, followed by spread to Central China, South Asia and Mesopotamia, before extension to Africa and the Mediterranean. Locations of domestication in South Asia and North China are thus rejected.

Access to the complete genome sequence makes it possible to estimate divergence times between populations analyzed by various evolutionary genetics approaches, using ***coalescent analysis*** and mutation dynamics in the genome. The application of these approaches, which had a pre-existing theoretical basis, has benefited from the exhaustive data provided by genome sequencing. The most recent studies (Lawal et al. 2020; Wang et al. 2020) indicate a divergence between wild and domestic populations dating from around 8,000 years BP (7,014–8,768 years ago) (Lawal et al. 2020) or 9,500 ± 3,300 years BP (Wang et al. 2020), without specifying the number of domestication centers. Based on the study by Peters et al.

(2022), the hypothesis of a single center in Southeast Asia seems the most likely. Wang et al. (2020) propose the subspecies *G. g. spadiceus* as the ancestral population. Arguments supporting an initial domestication from *G. g. spadiceus* are based, on the one hand, on the greater genetic proximity between this subspecies and Asian domestic chickens and, on the other hand, on the very high frequency in *G. g. spadiceus* of the alternative ***allele*** of the *TSHR* gene (see section 5.4.2.2), which is virtually fixed in modern domestic chickens (Rubin et al. 2010; Qanbari et al. 2019). However, the existence of subsequent exchanges between domestic chickens and other *G. gallus* subspecies is not ruled out by Wang et al. (2020).

5.3.2. *Guinea-fowl*

The guinea fowl, *Numida meleagris* (family *Numididae*), is native to Africa. It is the only species in the genus *Numida*, and its range covers a large part of sub-Saharan Africa, from Senegal to Eritrea and from Chad to South Africa, where eight subspecies have been identified. This species was probably domesticated around 2,000 years ago in northern sub-Saharan Africa. For a long time, the lack of ***molecular markers*** severely limited genetic studies concerning this species. A few studies have described the genetic diversity of domesticated and wild populations using microsatellites or mitochondrial DNA.

Recently, obtaining the entire guinea fowl genome sequence has enabled us to better understand the genetic structure of the species and the genetic diversity present in wild and domesticated populations. Studies point to an initial domestication in the West African zone, where wild populations in Burkina Faso (Vignal et al. 2019) or Nigeria (Shen et al. 2021) are genetically closer to all domestic populations than wild populations in other regions of Africa. Based on their data analyzing genetic structure, gene flow and demographic evolution, Shen et al. (2021) propose the existence of a single initial domestication event in West Africa. This event would have taken place between 5,500 and 1,300 years BP and corresponds to archaeological data. In both studies, domesticated populations are genetically closer to each other than to wild populations living in the same geographical area, although Shen et al. (2021) believe that there has been gene flow between wild and domesticated populations, at least in the Sudan region. This suggests that there is little exchange between wild and domestic populations in Africa, although village guinea fowl breeding is completely free-range. The arrival of domestic guinea fowl in Europe dates back to Greco-Roman times (2,000 years BP), after which they disappeared from Europe, only to be reintroduced in the 16th century by the Portuguese.

5.3.3. *Turkey*

The turkey, *Meleagris gallopavo* (family *Phasianidae*), is native to North America, from the eastern and southwestern United States to central Mexico. Six subspecies are recognized, with *M. g. sylvestris*, *M. g. merriami* and *M. g. osceola* (USA), *M. g. intermedia* (USA and Mexico), *M. g. mexicana* and *M. g. gallopavo* (Mexico), the last of which is now considered extinct. A second species, *M. ocellata*, is found in Mexico (Yucatan Peninsula), as well as northern Belize and Guatemala. The latter has not been domesticated. Archaeological data reveal two centers of domestication, one in the southwestern United States and the other in central Mexico, around 2,000 years ago (for a review, see Tixier-Boichard and Weigend (2020)). However, present day domestic turkey populations all seem to have descended from Mexican domestication, as shown by studies based on archaeological data and mitochondrial DNA polymorphism. In particular, Padilla-Jacobo et al. (2018) highlight the existence of three major mitochondrial groups: the first group is made up of individuals from the *M. g. mexicana* subspecies, the second group includes individuals from the *M. g. merriami* subspecies and the third group is a composite collection of subspecies (*M. g. mexicana/intermedia/silvestris/osceola*). All the domestic ***haplotypes*** studied belonged to this third group. Although genomic tools have been developed for this species in recent years, their application has mainly involved diversity analysis and the search for selection signatures, and has not studied domestication scenarios.

5.3.4. *Quail*

The Japanese quail, *Coturnix japonica* (family *Phasianidae*), is native to eastern Asia (Japan, China, Mongolia and Korea). The genus *Coturnix* comprises five other species, including *C. coturnix* (corn quail), which is found from central Siberia to Africa, through western and southern Europe. Both species are apparently *sympatric* from northern Mongolia to southern Lake Baikal. Only the *C. japonica* species has been domesticated, and it is therefore at the origin of all farmed and laboratory quail populations worldwide (for a review, see Tixier-Boichard and Weigend (2020)). The question of the geographical origin of domestication remains unanswered to this day, with the literature alternately supporting a Chinese origin (around 1,000–1,500 years ago) before its spread into Japan; and, through there, across the globe, or a Japanese origin (800 years ago) and its spread around the world. These studies have been carried out using microsatellite data and/or mitochondrial DNA sequences, and have also demonstrated the existence of gene flow in Europe between Japanese quail (from domestic animals released into the wild) and common quail, which can now be identified using an ***SNP panel***.

Knowledge of the complete quail genome is relatively recent. A few genomic studies of Japanese quail have been carried out recently (Morris et al. 2020). While they have revealed selection signatures (***selective sweeps***), they provide no new information on the geographical origin of the domestication of this species.

5.3.5. *Common duck*

The mallard, *Anas platyrhynchos* (family *Anatidae*), is present over a wide area, covering Europe, Asia and North America. The species is thought to have been domesticated in East Asia around 2,500 years ago and in Europe around 1,000 years ago. Various studies, based on microsatellite polymorphism or mitochondrial DNA, identify the mallard as the ancestral wild species, although some propose a double origin, with the mallard and the Chinese spot-billed duck, *A. zonorhyncha*, the two species being genetically very close. The genus *Anas* is quite diverse and includes 29 other species.

The first assembly of the common duck genome sequence was produced in 2013. Three studies were carried out using genomic data from domesticated and wild populations. Two studies involved the analysis of complete sequences of domestic individuals that solely belonged to Chinese populations or to Chinese and South and Southeast Asian populations, as well as wild individuals of the *A. platyrhynchos* species, sampled in China. They tend to support an origin of domestication in China, around 2,200 years ago. The third study combines sequences from Chinese domestic ducks and wild individuals of the two species *A. platyrhynchos* and *A. zonorhyncha* (Guo et al. 2021). It shows that these two species are genetically very close to one another, and that there is a complex and continuous gene flow between wild and domestic populations. The authors suggest that Chinese domestic populations may not derive from these two candidate species, but from another, currently unstudied, population. This highlights the need for these genomic studies to sample other populations of wild and domestic ducks outside the China/Southeast Asia area. The existence of data from ancient samples could also be a decisive factor in better understanding the domestication process of this species.

5.3.6. *Pigeon*

The rock dove, *Columba livia* (family *Columbidae*), is naturally present in Europe, North Africa, the Middle East and South Asia. It belongs to a highly diversified genus that includes 35 other species. Archaeological data suggest that domestication took place in Mediterranean and Middle Eastern regions between 3,000 and 5,000 years ago. Its domestication was accompanied at a very early stage by the development of numerous ***feral*** populations throughout the world. The first molecular genetic studies focused on the genetic structure (microsatellites or mitochondrial DNA) of domestic and feral populations of the species (Tixier-Boichard and Weigend 2020), but none specifically concerned domestication.

The huge morphological and behavioral diversity that has followed pigeon domestication (over 350 recognized breeds in Europe alone) has led to a great deal of work in genomics to understand the genetic basis of this diversity. The first genome sequence assembly for this species dates back to 2013, and another was proposed in 2018 (Holt et al. 2018). Applications have involved researching for selection signatures and identifying the molecular basis of phenotypic variations in plumage (color, structure, distribution), or flight ability and orientation in space. Several studies involve the genetic structure of domestic and feral populations.

5.4. Genetic mechanisms involved in domestication

The two approaches used to study these mechanisms differ according to whether or not they use phenotypic data. The joint analysis of molecular data and phenotypes is carried out on dedicated experimental setups, while the detection of molecular signatures without phenotypic data is based on genome sequencing of a panel of populations.

5.4.1. *Setups integrating phenotypic data*

5.4.1.1. *Experimental selection*

A divergent selection experiment of two lines was conducted in a captive population of red junglefowl on a measure (scored from 1 to 5) of fear of a person approaching a chicken in an enclosed space of 3 m^2. The ***heritability*** of this measure of "fear of man" was estimated at 0.17 at the start of the experiment. These two lines were compared on various traits at the eighth generation and were used for the analysis of gene expression profiles. The main results were summarized by Agnvall et al. (2018). They show that the less human-fearing line (L) has faster growth and a higher basal metabolism than the more human-fearing line (H), better feed efficiency (in females), increased serotonin (in males), and better plumage quality with the appearance of white spots on some animals. In terms of behavior, animals from the less fearful line were more aggressive and dominant toward those from the more fearful line, while no difference was observed between animals from the same line. No differences were observed in other behaviors, such as environmental exploration or dust bathing. Lastly, the less fearful line had a relatively smaller brain relative to body weight, but a relatively larger cerebellum. The relative weights of heart, liver, spleen and testes were also reduced in relation to body weight. Expression of thalamus and hypothalamus genes was compared between lines at generation 5. The results showed differences for genes associated with spermatogenesis and immune functions, but very few for behavior. This suggests that selection on taming has important effects on biological functions other than behavior, such as reproduction and immunity. However, no "key" or "conductor" gene could be associated with the phenotypic divergence obtained between the divergent lines. The authors therefore questioned the possible intervention of epigenetic mechanisms and identified differentially methylated regions (DMRs) between the divergent lines, which were not colocalized with the differentially expressed genes, suggesting distant regulatory effects.

5.4.1.2. *Comparing red junglefowl to a domestic chicken*

A line of domestic White Leghorn (WL) chickens was compared with a line of red junglefowl (RJ), sampled in Thailand. Phenotypic analysis revealed that the WL had both a higher body weight and a higher brain mass than RJ, but that brain mass to body weight ratio was lower in WL, with the exception of the cerebellum, as had been observed in the L and M lines. Expression of a pre-established list of candidate genes was also compared between WL and RJ, in five different tissues (hippocampus, hypothalamus, pituitary, adrenal glands and liver), at rest or 45 min after restraint stress. The results highlighted stress-related differences in gene expression, with overexpression of genes such as *FKBP5*, *C-FOS* or *EGR1* in the hippocampus and hypothalamus, and *StAR*, *MC2R*, *TH* in the adrenals. Comparison between lines showed that the *CRHR1* gene encoding the hormone receptor controlling corticosterone production was less expressed in the hippocampus and more expressed in the hypothalamus of WL. The CRH pathway is involved in fear memory. In the hypothalamus, genes encoding the glucocorticoid receptor or vasopressin were also overexpressed. These data are interpreted by the authors as indicating a stronger negative feedback on the hypothalamic-pituitary-adrenal axis, leading to a lower stress response in the domestic chicken. This assumes that the WL line is representative of all domestic chickens, which has not been demonstrated.

5.4.1.3. *QTL detection devices*

Identification of the *loci* determining differences in performance (quantitative traits or quantitative trait locus (***QTL***)) between populations or divergent lines can be achieved by cross-breeding to produce an *F1* generation, whose individuals are in turn cross-bred to produce an F2 generation. Further crossbreeding of individuals from each *F2* generation, then F3, then F4, etc., leads to the creation of a population known as Advanced InterCross (AIC), which improves the accuracy of detection of genes responsible for differences between parental lines (Figure 5.3). AIC populations have been created from generation 8 of divergent lines L and H, producing a population LH-F3, and from lines WL and RJ, up to generation AIC-F12.

The detailed description of behavior in the LH-F3 population confirmed that selection on taming had resulted in major behavioral changes that are characteristic of domestic chickens. The least fearful animals in LH-F3 also grew better through puberty. The distribution of the "fear of man" trait in LH-F3 was compatible with a polygenic genetic determinism.

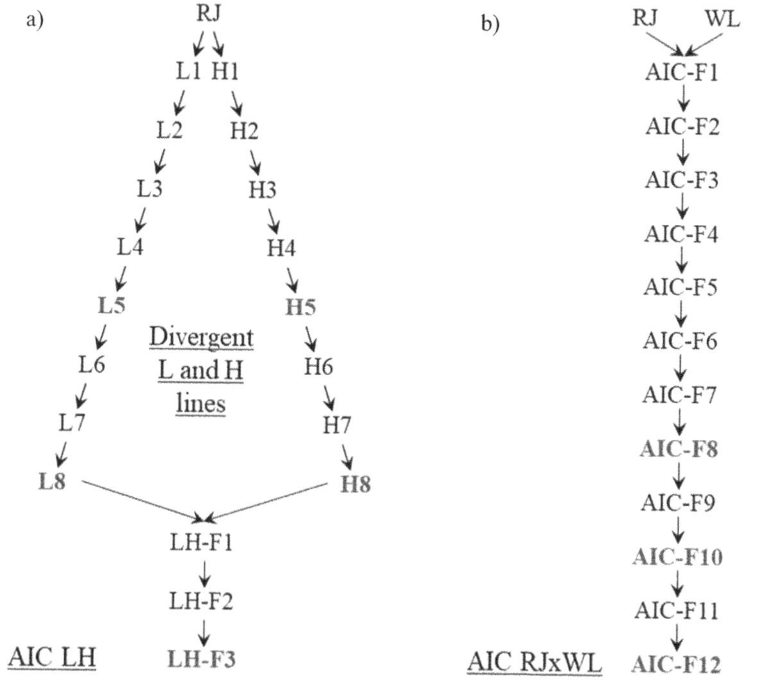

Figure 5.3. *Diagram of divergent lines and Advanced InterCrosses (AIC). For a color version of this figure, see www.iste.co.uk/pelletier/genetics.zip*

COMMENT ON FIGURE 5.3.– *(a) L and H lines divergently selected on "fear of man" (L for less fear, H for more fear) from a red junglefowl (RJ) population, with AIC-LH intercross from L8 and H8 populations. The generations described in the text (L5/H5, L8/H8 and LH-F3) are shown in bold and blue; (b) advanced intercross between populations of red junglefowl (RJ) and domestic White Leghorn (WL) chickens. The generations described in the text (AIC-F8, F10 and F12) are shown in bold and blue.*

The F8 generation from the RJxWL crossbreeding was characterized with phenotypic, genetic (with sparse genotyping counting fewer than 700 SNP markers) and ***transcriptomic*** analysis, with mapping of QTL controlling brain development (Henriksen et al. 2016), through to the identification of gene expression QTL (eQTL) on behavioral (Johnsson et al. 2018) and reproductive traits (Johnsson et al. 2016). The results showed that differences in body weight and brain weight between the WL and RJ lines were not controlled by the same genomic regions, leading to the conclusion that domestication specifically affects brain development. The AIC-F12 generation was particularly studied for its response to stress, with QTL and eQTL research on plasma levels of several hormones (corticosterone, dehydroepiandrosterone, aldosterone and pregnenolone). The *SERPINA10* gene was associated with variation in corticosterone levels and the *PDE1C* gene with aldosterone levels.

5.4.2. *A priori* molecular signature detection

5.4.2.1. *Principle*

Access to sequencing data has led to a growing number of studies comparing genome variability between domesticated and wild populations. The presence of current populations of the chicken's wild ancestor in its natural habitat is an advantage for this type of study, provided it is ensured that there has been no uncontrolled cross-breeding between wild and domesticated chickens, as likely happened to *Gallus sonneratii* animals in zoos around 150 years ago (Mariadassou et al. 2021). The principle behind the detection of molecular signatures is based on the very strong loss of variability in regions, subject to selection: if this loss of variability is observed between modern chickens and wild populations, with great homogeneity in the region identified between all modern chickens, but with significant variability observed in wild animals, it is possible to conclude that a domestication signature has been found. However, this conclusion depends on the completeness of the sampling and therefore needs to be supported by several independent studies. The ability to study ***ancient DNA*** samples adds a valuable time dimension, providing access to ancestors from archaeological sites or domesticated animals from antiquity or the Middle Ages. The study of ancient domesticated animals can help distinguish a domestication signature, which should be found in all domesticated chickens at all times, from a selection signature, which is only found in recent populations or in intensely selected domesticated populations. It should be noted that several genomic studies that led to the identification of domestication signatures only included intensely selected modern chickens (Rubin et al. 2010; Qanbari et al. 2019), unlike local breeds that are little subject to artificial selection, but subject to random drift effects. The study by Wang et al. (2020) included all population types, but the local breeds were mainly of Asian origin. Numerous signatures of domestication or selection have been published for chicken, but they vary in size and may correspond to several genes, whose role often remains hypothesized. A list of 23 genes emerges from the study by Rubin et al. (2010) and a list of 28 from that of Qanbari et al. (2019); only four of these are common to both studies.

This approach has also been used in guinea fowl (Vignal et al. 2019), with the identification of several candidate genes associated with important functions; in quail, with the identification of a single molecular signature that is common to quails selected for egg-laying or growth, thus possibly resulting from domestication; and in turkey, where only five regions of loss of diversity in domesticated birds are suggested, four of which are located at the end of chromosomes 9, 14 and 22, corresponding to homologous regions in chicken carrying growth QTLs. A recent analysis of the quail genome identified selection signatures on social motivation (Morris et al. 2020); however, none of the candidate genes mentioned correspond to those associated with fear of man in chickens, these two behaviors being completely different.

Two main results will be discussed in detail, before summarizing the genes derived from analyses with (or without) phenotypes.

5.4.2.2. TSHR *gene signature in chicken*

A key result of this approach was the identification of a strong genetic differentiation between red junglefowl and a panel of domestic lines of commercial origin (selected for egg-laying or growth) at the level of the TSH hormone receptor gene, which itself controls the production of T4 and T3 thyroid hormones. All domestic chickens carried the *TSHR* alternative allele, characterized by a ***non-synonymous mutation*** in a highly conserved region of the gene, although this allele was already present in some wild chickens (Rubin et al. 2010). However, analysis of DNA from domestic chickens dating back to Roman antiquity showed that the alternative *TSHR* allele was not fixed at that time, and that its frequency increased significantly in the early Middle Ages with the increasing presence of chicken farms, as attested by historical documents (Loog et al. 2017). The authors therefore concluded that the molecular signature observed on *TSHR* was a selection signature, rather than domestication. The functional effects of this allele were then analyzed in the F8 generation of the AIC RJxWL crossbred described above, where animals carrying the alternative "domestic" allele had a 4-h longer incubation time, less fearful behavior in males and less dominant behavior in females (Karlsson et al. 2015). The AIC-F10 generation was studied for egg-laying traits and sensitivity to an abrupt change in the day/night rhythm. Females homozygous for the alternative *TSHR* allele reached peak egg-laying faster than wild-type females, but overall egg-laying intensity did not differ between the two genotypes. Males homozygous for the alternative *TSHR* allele showed less variation in testicular weight and in the expression of thyroid axis genes (TSH hormone and its receptor, type-2 or type-3 deiodinases), which the authors interpreted as less sensitive to photoperiod than in the wild ancestor. This characteristic is indeed important in domestic chickens, as it enables them to lay eggs for longer. Nevertheless, the domestic chicken remains photosensitive, but intensely selected lines are much less photosensitive than local breeds. A recent functional genomics analysis in quail confirmed the effect of photoperiod on the regulation of genes encoding TSHβ and deiodinases (Morris et al. 2020).

We can therefore conclude that the "domestic" allele of the *TSHR* gene acts on functions that are critical to domestication (behavior and reproduction), and that its significance has been amplified by the selection that followed domestication. It should be noted that *TSHR* was not a candidate gene for domestication, which shows the additional contribution of unbiased analyses in relation to targeted functional studies.

5.4.2.3. *Results supporting the neural crest hypothesis*

Neural crest theory has been proposed to explain the biological mechanisms involved in mammalian domestication (Wilkins et al. 2014). Selection for increased docility is likely to involve neural changes, notably concerning hypothalamic-pituitary regulation of the adrenal system, which plays a key role in stress and behavior. At the same time, coat color changes occur, often with the appearance of white patches. Taken together, these findings suggest the involvement of the neural crest cell lineage, the origin of melanocytes as well as neurons, which contribute to brain development and neuroendocrine functions, including those influencing stress. Therefore, hypofunction of the neural crest would both reduce neural input to the sympathetic and adrenal systems and explain pigmentation changes.

Evidence supporting this theory in birds was recently obtained in guinea fowl (Vignal et al. 2019), with the identification of a domestication signature on the *EDNRB2* gene, also known to induce a phenotype of white patches, or even extreme dilution of plumage color in chickens. Curiously, village guinea fowl studied in West Africa, in countries where wild guinea fowl had also been studied, showed areas of white skin on the head, compared with the black skin of their wild counterparts, with many of them either having extensive white patches or a paler, diluted color. Strikingly, alleles of the *EDNRB2* gene have also been associated with white patches or very diluted plumage in quail, duck, goose and pigeon, which tends to confirm the frequency of depigmentation in domestic birds. To fully support the neural crest hypothesis, it remains to be seen whether the behavior of animals with depigmented plumage differs from that of normally pigmented animals. This has not yet been done, as plumage color variants are generally fixed in standardized breeds, and a comparison system involves building up a dedicated informative population. Moreover, there are many breeds of chicken that are totally black or red, so the presence of white spots is not, or no longer, a distinctive sign of domestication; whereas, it is in guinea fowl, a more recently domesticated species.

Other genes involved in neural crest development or pigmentation have also been associated with a molecular signature distinguishing red junglefowl from commercial chicken lines: this is the case for *CORIN*, *SKIV2L2* and *KITLG* (Qanbari et al. 2019), *FGFR1*, *MYC-1*, *ERBB4*, and BMP family genes (Wang et al. 2020).

5.4.3. *Integrating approaches*

Tables 5.2–5.5 summarize the candidate genes identified during QTL or eQTL detection studies and unbiased molecular signature analyses. It should already be noted that this list does not include any of the genes found to be differentially expressed between red junglefowl lines that are more or less fearful of humans (see section 5.4.1.1) or between a red junglefowl and a White Leghorn domestic chicken (see section 5.4.1.2). Furthermore, comparison of the results of the AIC RJxWL populations with those of whole-genome signature detection only shows three cases of common genes: the *TANK* and *SHANK2* genes, each associated with a QTL on cerebellum mass and found in a wild/domestic or wild/layer chicken selection signature, and the *TSHR* gene, first identified by selection signature, and then confirmed in generation 8 of the AIC RJxWL population.

Gene	Functional area or phenotype	Approach	Source
ASNS	Asparagine synthesis	(3)	(f)
STX2	Intracellular trafficking (SNARE binding)	(3)	(f)
EGFR	Cell proliferation	(3)	(f)
PLOD2	Cell bonding, collagen	(3)	(a)
OSGIN1	Oxidative stress	(3)	(a)
BRCA1	Transcription	(3)	(a)
NBR1	Cell proliferation	(3)	(a)
BAZ2B	Transcription, chromatin	(3)	(a)
TSNARE1	Intracellular traffic	(3)	(a)
EML4	Cytoskeleton	(3)	(a)
ECR3	Transcription	(3)	(a)
ANK2	Cytoskeleton	(3)	(a)
SLC16A12	Transmembrane transport	(3)	(a)

Table 5.2. *List of genes potentially associated with domestication, relating to cellular metabolism*

COMMENT ON TABLE 5.2.– *(1) QTL detection in AICs; (2) eQTL detection in AICs; (3) selection signatures; (4) frequency contrast for non-synonymous mutations between wild and domestic; (5) genotype comparison at generation 8 of the CBxWL AIC population. (a) Rubin et al. (2010); (b) Henriksen et al. (2016); (c) Johnsson et al. (2016); (d) Johnsson et al. (2018); (e) Wu et al. (2018); (f) Qanbari et al. (2019); (g) Vignal et al. (2019); (h) Wang et al. (2020); (i) Karlsson et al. (2015).*

Altogether, 13 genes (Table 5.2) were found to be involved in cell or nucleus metabolism (proliferation, cytoskeleton and ***transcription***) and 22 in growth, reproduction or immunity (Table 5.3). Genes known to be involved in embryonic development, and in particular nervous system development, are more numerous, with seven genes annotated in connection with the neural crest (Table 5.4) and 28 involved in development, notably of the brain, or in behavior (Table 5.5).

Gene	Functional area or phenotype	Approach	Source
FGF1	Fertility	(1) (2)	(c)
MRPL42	Fertility	(1) (2)	(c)
MRPL32	Fertility, hypothalamic regulation	(1) (2)	(c)
FGF1	Egg laying	(1)	(c)
MRPL42	Egg laying	(1)	(c)
MRPL32	Egg laying	(1)	(c)
GNRH-1	Reproduction	(3)	(h)
KIF18A	Reproduction	(3)	(h)
PAPA2	Growth (guinea fowl)	(3)	(g)
IGF1	Growth	(3)	(a) (f)
TBXAS1	Metabolism, cardiovascular disease	(3)	(f)
C1GALT1	Protein glycosylation	(3)	(f)
THSD4	Metabolism, peptidase activity	(3)	(f)
Gga-mir-1715	Immunity	(3)	(f)
BCO2	Carotenoid metabolism	(3)	(a) (f)
SPEF2	Spermatogenesis	(4)	(f)
LEPR	Appetite, body weight	(4)	(f)
IGSF10	Reproduction	(4)	(f)
PLEKHM1	Osteoporosis	(4)	(f)
MEGF10	Myogenesis	(4)	(f)
VSTM2A	Adipocyte development	(3)	(a)
LIPA	Metabolism, lipase	(3)	(a)

Table 5.3. *List of genes potentially associated with domestication, relating to production traits (same notations as in Table 5.2 for type of approach and source (see comment in Table 5.2))*

Gene	Functional area	Approach	Source
CORIN	Neural crest	(4)	(f)
EDNRB2	Neural crest (guinea fowl)	(3)	(g)
ERBB4	Neural crest	(3)	(h)
FGFR1	Neural crest	(3)	(h)
KITLG	Neural crest	(3)	(f)
MYC-1	Neural crest	(3)	(h)
SKIV2L2	Neural crest	(4)	(f)

Table 5.4. *List of genes potentially associated with domestication and associated with neural crest development (same notations as in Table 5.2 for type of approach and source (see comment in Table 5.2))*

It should be noted that QTL detection studies on AIC crosses have used very few markers: the position of the QTL is therefore not very precise. The cross between divergent lines of red junglefowl is very convincing in terms of behavioral phenotype analysis; the work clearly shows that the genetic determinism involved is polygenic in nature, and that significant differences appear in less than 10 generations of divergent selection. The cross between a red junglefowl and a domestic Leghorn chicken is not comparable, as the Leghorn chicken has also been strongly selected for egg-laying. The comparison does not just involve domestication, and the number of generations separating the two crossbred populations is much higher than between divergent lines of red junglefowl; therefore, random drift effects may also play a role. The advantage of the eQTL approach is that it avoids this hazard, as it is associated with proven functional information (differences in gene expression).

Furthermore, annotation of the chicken genome is not as good as that of humans or mice yet. Functional annotation of genes is still incomplete, which limits functional interpretations. As a result, the identification of candidate genes in large QTL regions remains highly imperfect. Similarly, in selection signatures, while some regions do show a significant gap in terms of diversity, pinpointing the gene(s) associated with this signature remains highly speculative. There is certainly a proportion of false positives, which is impossible to detect with the available data.

Gene	Functional area or phenotype	Approach	Source
ACOT9	Social behavior	(1)	(d)
AGTR1	Angiotensin, aldosterone regulation	(3)	(a) (f)
ALS2	Guanine metabolism, neurodegeneration	(3)	(a)
ALX1	Brain development	(3)	(f)
CAMK2D	Protein kinase activity	(3)	(f)
CCKBR	Cerebellum mass, synapses	(1)	(b)
CCNF	Cyclin, development	(3)	(f)
CHD5	Social behavior	(2)	(d)
COT02188	Development (quail)	(3)	(e)
CRAMP1L	Development	(3)	(f)
FGF3, FGF9	Cerebellum development	(1)	(b)
GLI3	Wing development	(4)	(f)
GLP1R	Aldosterone synthesis, glucagon receptor	(3)	(f)
HNF4G	Development, transcription	(3)	(a) (f)
JPT2	Development	(3)	(f)
KIF7	Wing development	(4)	(f)
MAP6	Cerebellum mass	(1)	(b)
MPP4	Intercellular junctions, retina	(3)	(a)
PDE1C	Stress	(1) (2)	(b)
PLEKHB1	Cerebellum mass	(1)	(b)
PRDX4	Social behavior	(1)	(d)
SEMA3D	Axon development	(3)	(a)
SERPINA10	Stress	(1) (2)	(b)
SHANK2	Cerebellum, synapses	(1) (3)	(a) (b)
SRPX	Social behavior	(1)	(d)
TANK	Cell signaling, cerebellum mass	(1) (3)	(a) (b)
TSHR	Photoperiodism, thyroid	(3) (4) (5)	(a) (f) (i)
TTRAP	Social behavior	(1) (2)	(d)

Table 5.5. *List of genes potentially associated with domestication in relation to embryonic development, the nervous system or behavior (same notations as in Table 5.2 for type of approach and source (see commentary in Table 5.2))*

Subsequently, functional studies indicate changes that may be under the influence of distant regulatory elements that are still poorly understood. As a result, the discussion of selection signature results focuses on regions with known genes.

Ultimately, confusion between selection and domestication signatures is certainly common, but domestication can be considered a continuous process and there is no clear dividing line between a "domesticated" and "selection" state, as shown by the case of the alternative *TSHR* allele.

5.5. Conclusion

The domestication of the chicken most certainly involved a taming effect by man, which modified the genetic architecture of the bird's brain development and rapidly led to the first observed changes in growth. It may be tempting to extend this interpretation to other domesticated birds, but there are no data to support it. Access to current populations of the wild ancestor is a definite advantage for this behavioral approach, which would be interesting to test out in other bird species of breeding interest. The suggested link between behavior, neural crest development and depigmentation phenotype in mammals is also being confirmed in birds. Over time, the frequency of the alternative *TSHR* allele, or that of the *BCO2* "yellow skin" allele, has increased according to the importance breeders have tied to them in terms of the phenotypic consequences of these variants.

There is no clear dividing line between the domestication and selection phases: the processes are continuous, which is why unbiased molecular signature studies detect the consequences of both phases. However, as the selection of farm animals has intensified over the last century, its impact on the genome is becoming more dominant. The possibility of studying the wild ancestor, that is still present and accessible, even under experimental conditions, is a major asset for understanding the initial process of domestication in birds.

5.6. References

Agnvall, B., Belteky, J., Katajamaa, R., Jensen, P. (2018). Is evolution of domestication driven by tameness? A selective review with focus on chickens. *Applied Animal Behaviour Science*, 205, 227–233.

Barton, L., Bingham, B., Sankaranarayanan, K., Monroe, C., Thomas, A., Kemp, B.M. (2020). The earliest farmers of northwest China exploited grain-fed pheasants not chickens. *Scientific Reports*, 10(1), 2556.

Guo, X., He, X.X., Chen, H., Wang, Z.C., Li, H.F., Wang, J.X., Wang, M.S., Jiang, R.S. (2021). Revisiting the evolutionary history of domestic and wild ducks based on genomic analyses. *Zoological Research*, 42(1), 43–50.

Henriksen, R., Johnsson, M., Andersson, L., Jensen, P., Wright, D. (2016). The domesticated brain: Genetics of brain mass and brain structure in an avian species. *Scientific Reports*, 6, 34031.

Holt, C., Campbell, M., Keays, D.A., Edelman, N., Kapusta, A., Maclary, E., Domyan, E., Suh, A., Warren, W.C., Yandell, M. et al. (2018). Improved genome assembly and annotation for the rock pigeon (*Columba livia*). *G3 Genes Genomes Genetics*, 8(5), 1391–1398.

Johnsson, M., Jonsson, K.B., Andersson, L., Jensen, P., Wright, D. (2016). Quantitative trait locus and genetical genomics analysis identifies putatively causal genes for fecundity and brooding in the chicken. *G3 Genes Genomes Genetics*, 6(2), 311–319.

Johnsson, M., Henriksen, R., Fogelholm, J., Hoglund, A., Jensen, P., Wright, D. (2018). Genetics and genomics of social behavior in a chicken model. *Genetics*, 209(1), 209–221.

Karlsson, A.C., Svemer, F., Eriksson, J., Darras, V.M., Andersson, L., Jensen, P. (2015). The effect of a mutation in the thyroid stimulating hormone receptor (TSHR) on development, behaviour and TH levels in domesticated chickens. *PLoS ONE*, 10(6), e0129040.

Kuhl, H., Frankl-Vilches, C., Bakker, A., Mayr, G., Nikolaus, G., Boerno, S.T., Klages, S., Timmermann, B., Gahr, M. (2021). An unbiased molecular approach using 3'-UTRs resolves the avian family-level tree of life. *Molecular Biology and Evolution*, 38(1), 108–127.

Lawal, R.A., Martin, S.H., Vanmechelen, K., Vereijken, A., Silva, P., Al-Atiyat, R.M., Aljumaah, R.S., Mwacharo, J.M., Wu, D.D., Zhang, Y.P. et al. (2020). The wild species genome ancestry of domestic chickens. *BMC Biology*, 18(1), 13.

Loog, L., Thomas, M.G., Barnett, R., Allen, R., Sykes, N., Paxinos, P.D., Lebrasseur, O., Dobney, K., Peters, J., Manica, A. et al. (2017). Inferring allele frequency trajectories from ancient DNA indicates that selection on a chicken gene coincided with changes in medieval husbandry practices. *Molecular Biology and Evolution*, 34(8), 1981–1990.

Mariadassou, M., Suez, M., Sathyakumar, S., Vignal, A., Arca, M., Nicolas, P., Faraut, T., Esquerré, D., Nishibori, M., Vieaud, A. et al. (2021). Unraveling the history of the genus Gallus through whole genome sequencing. *Molecular Phylogenetics and Evolution*, 158, 107044.

Morris, K.M., Hindle, M.M., Boitard, S., Burt, D.W., Danner, A.F., Eory, L., Forrest, H.L., Gourichon, D., Gros, J., Hillier, L.W. et al. (2020). The quail genome: Insights into social behaviour, seasonal biology and infectious disease response. *BMC Biology*, 18(1), 14.

Padilla-Jacobo, G., Cano-Camacho, H., López-Zavala, R., Cornejo-Pérez, M.E., Zavala-Páramo, M.G. (2018). Evolutionary history of Mexican domesticated and wild Meleagris gallopavo. *Genetics Selection Evolution*, 50(1), 19.

Peters, J., Lebrasseur, O., Irving-Pease Evan, K., Paxinos Ptolemaios, D., Best, J., Smallman, R., Callou, C., Gardeisen, A., Trixl, S., Frantz, L. et al. (2022). The biocultural origins and dispersal of domestic chickens. *Proceedings of the National Academy of Sciences*, 119(24), e2121978119.

Qanbari, S., Rubin, C.J., Maqbool, K., Weigend, S., Weigend, A., Geibel, J., Kerje, S., Wurmser, C., Peterson, A.T., Brisbin, I.L., Jr. et al. (2019). Genetics of adaptation in modern chicken. *PLoS Genetics*, 15(4), e1007989.

Rubin, C.J., Zody, M.C., Eriksson, J., Meadows, J.R., Sherwood, E., Webster, M.T., Jiang, L., Ingman, M., Sharpe, T., Ka, S. et al. (2010). Whole-genome resequencing reveals loci under selection during chicken domestication. *Nature*, 464(7288), 587–591.

Shen, Q.K., Peng, M.S., Adeola, A.C., Kui, L., Duan, S., Miao, Y.W., Eltayeb, N.M., Lichoti, J.K., Otecko, N.O., Strillacci, M.G. et al. (2021). Genomic analyses unveil helmeted guinea fowl (*Numida meleagris*) domestication in West Africa. *Genome Biology and Evolution*, 13(6), evab090.

Tixier-Boichard, M. and Weigend, S. (2020). The origin and domestication of poultry species. In *Advances in Poultry Genetics and Genomics*, Aggrey, S.E., Tixier-Boichard, M., Rhoads, D.D. (eds). Burleigh Dodds Science Publishing, Cambridge.

Vignal, A., Boitard, S., Thébault, N., Dayo, G.K., Yapi-Gnaore, V., Youssao Abdou Karim, I., Berthouly-Salazar, C., Pálinkás-Bodzsár, N., Guémené, D., Thibaud-Nissen, F. et al. (2019). A guinea fowl genome assembly provides new evidence on evolution following domestication and selection in galliformes. *Molecular Ecology Resources*, 19(4), 997–1014.

Wang, M.S., Thakur, M., Peng, M.S., Jiang, Y., Frantz, L.A.F., Li, M., Zhang, J.J., Wang, S., Peters, J., Otecko, N.O. et al. (2020). 863 genomes reveal the origin and domestication of chicken. *Cell Research*, 30(8), 693–701.

Wilkins, A.S., Wrangham, R.W., Fitch, W.T. (2014). The "domestication syndrome" in mammals: A unified explanation based on neural crest cell behavior and genetics. *Genetics*, 197(3), 795–808.

Wu, Y., Zhang, Y., Hou, Z., Fan, G., Pi, J., Sun, S., Chen, J., Liu, H., Du, X., Shen, J. et al. (2018). Population genomic data reveal genes related to important traits of quail. *GigaScience*, 7(5), 1–16.

6

Genetics of Fish Domestication in Aquaculture

Fabrice TELETCHEA
URAFPA, Université de Lorraine, Nancy, France

6.1. Introduction

During the domestication of fish, these animals gradually diverged from their wild ancestors, and are no longer considered captive wild animals, or even tame, but domesticated. Nevertheless, it is important to stress that the distinction between wild and domestic animals remains blurred, and we must not allow wild/domestic to represent a dichotomy such as black/white or true/false. On the contrary, there is a continuum between these two extremes, and considering a species as domesticated is somewhat arbitrary and questionable for fishes, which have only undergone a recent domestication, with the exception of a handful of species (Teletchea 2016, 2019). As a result, drawing up a list of domesticated fish in aquaculture today can be a tricky, even unrealistic task, in the absence of consensus on the very definition of a domesticated species (Balon 2004; Bilio 2008; Teletchea 2012).

To get around this problem, while still trying to describe the diversity of domestication histories in fish as well as possible, and also being able to compare them with one another, we proposed the concept of domestication levels around 10 years ago (Teletchea 2012). The aim of this concept is to account for the degree of mastery of the lifecycle of fish in captivity and the degree of freedom from wild populations (Figure 6.1). Five successive levels of domestication have been

Genetics of Domestications,
coordinated by Georges PELLETIER.

proposed: initial breeding trials (level 1), controlling part of the lifecycle in captivity (level 2), controlling the entire lifecycle in captivity (level 3), controlling the lifecycle without wild input (level 4) and implementation of breeding programs (level 5). Although this concept was originally developed for fish, it can also be applied to other zoological groups, such as zoo mammals.

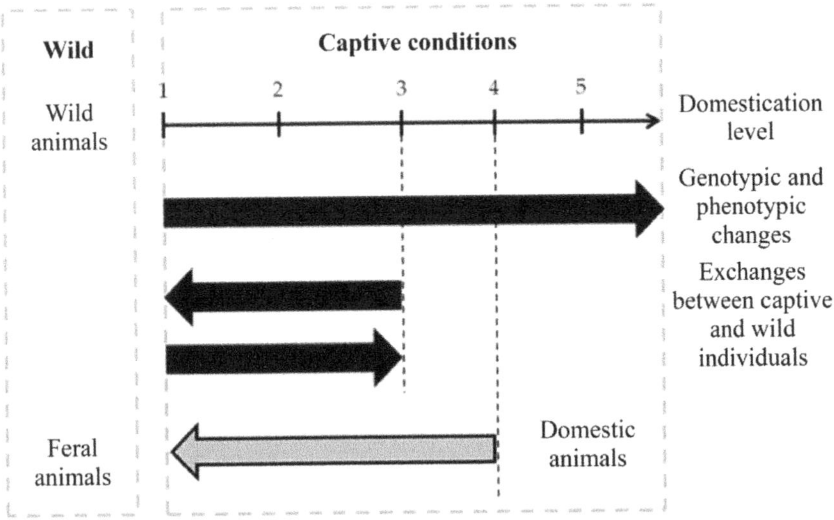

Figure 6.1. *Evolution of an animal species during the domestication process: 1–5 represents levels of domestication (source: F. Teletchea)*

As shown in Figure 6.1, domestication begins when a group of wild fish is transferred from its natural environment to captive conditions (level 1). Five genetic processes can be considered to be involved in the subsequent evolution of these animals (Teletchea 2021). From the outset, these include two uncontrolled processes, inbreeding and genetic drift, which result from the small size of founder populations, causing random changes in initial *allele* frequencies. Second, two partially controlled processes are involved: the natural selection imposed by breeding conditions on captive populations, which is not due to active selection (and only involved from level 5 upwards), and the release of previous natural selective pressures that can accompany the transition from the wild to the captive environment.

The first partially controlled process leads to the elimination of animals that are unable to survive and reproduce in captivity; conversely, animals that can produce a large number of offspring in the human-made environment become the majority, and ultimately the only ones present in farms. In the absence of active selection (level 5), natural selection provides the basic selective mechanism for genetic change in captive populations. The intensity of this natural selection in captivity depends, on the one hand, on the potential that the captive environment offers to develop and express the biological traits typical of the species, and, on the other hand, on the number of generations in captivity. Given that most species have relatively few pre-adaptations to captivity, natural selection is most intense in the first few years following the transition from the wild environment to the captive environment (essentially at levels 1–3).

The second partially controlled process, the relaxation of natural selection, involves a reduction in selective pressure. Certain behaviors that are important for survival in the wild, but not in captivity, such as foraging and anti-predator behavior, as well as certain morphological traits (coloration patterns, for example), can progressively be lost in captivity. Additionally, the genetic and phenotypic variability of these traits may become more significant as domestication progresses.

Lastly, the fifth genetic process is controlled, and consists of active selection, as changes are oriented (level 5). Active selection, which is the only mechanism that is unique to domestication, implies that man selects the animals that will participate in reproduction and enables the creation of different desired phenotypes (such as better growth, particular morphology, disease resistance and fillet yield).

During domestication, fish evolve from a genetic and phenotypic point of view, particularly when exchanges with wild conspecifics no longer exist (level 4). It is important to note that domesticated animals (levels 4 and 5) can return to the wild and are then referred to as feral animals. The differences between wild and feral animals will mainly depend on the time spent in captivity, and particularly on the number of generations there have been without the exchange of individuals between domesticated animals and their wild conspecifics (Teletchea 2019).

6.2. Diverse, complex and poorly understood domestication histories

The objective in aquaculture is generally to reach level 4 or 5 as quickly as possible, in other words, to fully control the fish's lifecycle with no input from the wild (Bilio 2008). This means regulating the rearing of eggs, larvae, juveniles and broodstock (Figure 6.2). The main obstacles to this total control of the lifecycle (level 3 and beyond) are (i) inadequate feed, particularly for the first feeding of the tiny larvae of many fish species; (ii) poor development of reproductive organs due to unsuitable environmental conditions (temperature, light and water quality, in particular); and (iii) absence of spawning (no reproductive behaviors that allow gamete exchange). Once all these bottlenecks have been overcome, the domestication process can continue, with the aim of progressively improving animal performance (better survival, better growth) and breeding practices (improved feeding).

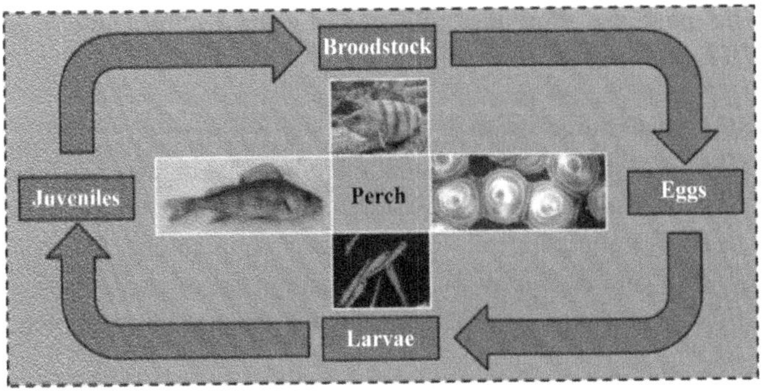

Figure 6.2. *Lifecycle of a fish, taking the perch as an example (Perca fluviatilis), which is a freshwater species of interest for European continental fish farming (source: Fontaine and Teletchea (2019)). For a color version of this figure, see www.iste.co.uk/pelletier/genetics.zip*

The first attempts to farm fish species for human consumption may date back more than 8,000 years, with the aquaculture of common carp (*Cyprinus carpio*) in Henan province, China (Nakajima et al. 2019). Traces of farming have also been found on representations in Egyptian tombs dating back around 3,500 years, with some control over the breeding of Nile tilapia (*Oreochromis niloticus*) in irrigation ponds (Teletchea 2021).

In Europe, pond farming of common carp was already developed in the Middle Ages. The Italian *Vallicoltura* (extensive farming of various marine species in coastal lagoons) was developed as early as the Middle Ages. Extensive farming of various marine species in coastal lagoons and large open lakes dates back to the 15th century. French trout farming began in the second half of the 19th century.

In North America, aquaculture began around 100 years ago. In Africa, aquaculture dates back to the 1940s. In Australia, New Zealand and various Pacific island states, the development of aquaculture is even more recent. In conclusion, although the first fish farms date back a long way, particularly in Asia, it was not until the early 1980s that aquaculture really took off, becoming the world's fastest growing food production sector (Teletchea 2016). Nevertheless, this production boom should not obscure the fact that it is likely that dozens, if not hundreds, of attempts to farm new fish species ended in very low production (a few dozen tons) or failure, with no written evidence of these. It is likely that these failures are linked, on the one hand, to a lack of sufficient biological, ecological and zootechnical knowledge of the target species, and, on the other hand, to excessively high production costs or low market acceptance of these new species (Bilio 2008).

As a result, three-quarters of current production is based on 10 species (Table 6.1), some of which, like common carp and Nile tilapia, have been introduced into a large number of countries. Conversely, of the 250 species farmed worldwide in recent decades, almost half have achieved an annual production of less than 100 tons.

Scientific name	Common name	Production	%	N
Ctenopharyngodon idellus	Grass carp	4,159,919	12.6	5
Hypophthalmichthys molitrix	Silver carp	4,075,115	12.3	5
Cyprinus carpio	Common carp	3,216,203	9.7	5
Oreochromis niloticus	Nile tilapia	2,542,960	7.7	5
Hypophthalmichthys nobilis	Bighead carp	2,466,578	7.4	5
Catla catla	Catla	2,418,821	7.3	5
Carassius carassius	Crucian carp	2,057,104	6.2	5
Salmo salar	Atlantic salmon	1,440,725	4.3	5
Labeo rohita	Roho labeo	1,221,828	3.7	5
Pangasius spp	Panga	1,193,023	3.6	5

Table 6.1. *Ranking of the 10 species whose production exceeded 1 million tons in 2009, with their contribution to world volume (%), and their level of domestication (N)*

The level of domestication also greatly varies from species to species. The 10 most widely produced species in the world have all reached level 5 (Table 6.1), while, in contrast, 70% of species have never reached level 4 (Figure 6.3).

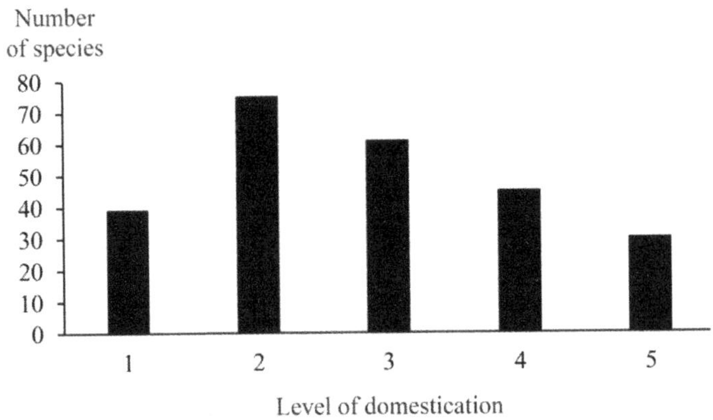

Figure 6.3. *Number of species by level of domestication (in 2009) (source: F. Teletchea)*

6.3. Significant performance improvements for domesticated species

Once a species has reached level 4, it is possible to implement breeding programs (level 5). Historically, these programs have focused on improving growth (Gjedrem et al. 2012). In Atlantic salmon, for example, the genetic gain in growth rate averaged 10–20% per generation over the first five generations of selection (Gjedrem et al. 2012). In addition to growth, food conversion efficiency, age of sexual maturity, improved resistance to bacterial and viral diseases, and a number of product quality-related traits (e.g. muscle lipid content, flesh color, tenderness and flavor) have been progressively included in various breeding programs (FAO 2019; Houston et al. 2020). In a recent survey conducted among selective breeding companies for five species bred in Europe, it was found that growth performance was subject to universal selection (Janssen et al. 2017). Among the 27 selection programs, morphology and disease resistance were included in 15 programs, product quality in 13 programs, processing yield in 12 programs, and reproduction and food efficiency in seven programs (Janssen et al. 2017).

This considerable improvement in fish performance in aquaculture over the last 40 years or so has been made possible by the tremendous progress in genetic selection techniques and concepts (Gjedrem and Baranski 2009). Initially developed for farmed terrestrial animals, these techniques were first applied to the 10 or so fishes with the greatest economic value (Table 6.1), then progressively applied to a growing number of species, thanks in particular to the reduction in the cost of analyses and the democratization of selection methods (Gjedrem and Baranski 2009;

Houston et al. 2020). The oldest and simplest selection method is solely based on individual performance, also known as massal selection (Gjedrem and Baranski 2009; Vandeputte 2009). It consists of selecting individuals, from among the captive population, that have the best performance (e.g. the best growth rate). This method therefore requires no a priori knowledge of kinship between fish. Nevertheless, it can only be applied if the trait can be measured without sacrificing the animal, and will be all the more effective if the selected trait has a high ***heritability*** (in other words, its phenotypic expression essentially depends on the genetic heritage of the individual and is only slightly influenced by environmental conditions). This method is still the most widely used in the absence of any means of tracking an individual's genealogy.

In cases where it is possible to establish precise kinship links between fish, other selection methods, known as family selection, can be used, such as those based on the performance of collaterals (brothers and sisters, half-brothers and half-sisters), ascendants (parents, grandparents) or descendants (children, grandchildren, etc.) (Vandeputte 2009). These are very interesting methods for measuring performances that require sacrificing the animal, such as meat quality, or that are not easily analyzed under rearing conditions, such as disease resistance (Gjedrem and Baranski 2009). Taking the performance of related individuals into account also considerably increases the efficiency of mass selection. Nevertheless, family breeding programs are heavy and costly in terms of infrastructure and personnel, and are only economically feasible for species with sufficient production levels, as is the case for salmonids in Norway, the United States, Finland and Chile (Vandeputte 2009).

In parallel with selection methods that are solely based on phenotyping (what can be seen or measured on the animal or its relatives), new methods directly analyzing all or part of the fish genome have improved the efficiency of traditional methods (Vandeputte 2009; Houston et al. 2020). The analysis of various genetic markers, ***microsatellites*** and ***SNPs*** has enabled the development of marker-assisted selection (MAS) methods (Vandeputte 2009). Implementing MAS requires rigorous organization of selection programs (individual marking, DNA collection from fish, genotyping) (Vandeputte 2009). In addition, advances in genetics now make it possible to directly select individuals with the best genotype, for example, favorable alleles of one or more ***QTLs***, even if this type of selection is complex to implement (Vandeputte 2009).

One of the best-known examples is the identification of a QTL on chromosome 26, involved in resistance to the virus responsible for infectious pancreatic necrosis (IPN) in Atlantic salmon, which could cause up to 90% mortality in farms

(Houston et al. 2020). Following the selection of resistant individuals, mortality became almost zero, although the precise mechanisms of resistance remain poorly understood (Houston et al. 2020). In Atlantic salmon again, another ***locus***, *vgll3*, has been identified and explains 30–40% of the phenotypic variation in the age of sexual maturation (Houston et al. 2020).

More recently, ***genomic selection***, which was initially developed for terrestrial livestock, and dairy cattle in particular, has been applied to aquaculture breeding. This involves comparing the genotype and phenotype of a reference population in order to establish a predictive model, which is then applied to genotyped individuals that are candidates for selection (Houston et al. 2020). Over the past five years, genomic selection has been applied to more than a dozen fish species, most often to improve growth and disease resistance (Houston et al. 2020).

In conclusion, the aquaculture industry will most probably continue to focus primarily on improving the growth rates of fish species that have already been domesticated, while simultaneously adding other traits, notably resistance to diseases. It is also very likely that genetic techniques and methods will continue to expand within the aquaculture industry, at all levels of domestication, so as to help improve fish performance while maintaining sufficient genetic variability across generations.

6.4. A success story: Atlantic salmon

Atlantic salmon farming began in the early 19th century in the UK, with the goal of replenishing river populations for angling (Teletchea 2016). It was not until the late 1960s that sea cage farming was first used in Norway. Starting in 1970, the very first family-based Atlantic salmon farming program was launched in Norway at a brand-new research institute called AKVAFORSK (Gjedrem 2012). Although Atlantic salmon production was only 100 tons in Norway at the time, the equivalent of around 3.6 million US dollars was invested in this program, two-thirds of this investment being by the Norwegian government and the rest by industry (not involved in aquaculture) and non-governmental organizations (Gjedrem 2012).

Meanwhile, research programs were launched to develop new dry feed pellets for Atlantic salmon, which became available in 1982. After around 40 years of farming, the time needed to produce a standard 4 kg fish has been halved, and while 3 kg of dry matter (in wet feed) was needed to produce 1 kg of salmon at the start, this figure has since been reduced to 1.15 kg of dry matter (Gjedrem 2012). Other traits have gradually been added to the selection objectives, such as age of sexual maturity, resistance to disease and stress, and flesh quality (Gjedrem et al. 2012). From the earliest stages of selective breeding programs, eggs and juveniles were sold to industry, so that almost 100% of salmon production in Norway and the rest of the world is now based on improved stocks (Gjedrem et al. 2012). Farmed salmon is considered one of the most domesticated fish species for human consumption; one Norwegian strain is currently more than 15 generations into breeding (Glover et al. 2017, 2020).

With domestication, production has increased 5,000-fold (Figure 6.4), from 294 tons in 1970 to over 1.4 million tons in 2010. A dozen countries currently produce salmon, the four largest being Norway, Chile, the UK and Canada. In Norway and Chile (where the species was introduced), salmon farming is of considerable economic importance. In Norway, salmon farming is the third largest industry after oil and mining. In Chile, salmon exports account for around two-thirds of total fishery exports and have become the third most significant export product after copper and wood products. Most production (grow-out) is now carried out in sea cage farming, with hatcheries (egg and small fish production) working with land-based recirculating aquaculture systems.

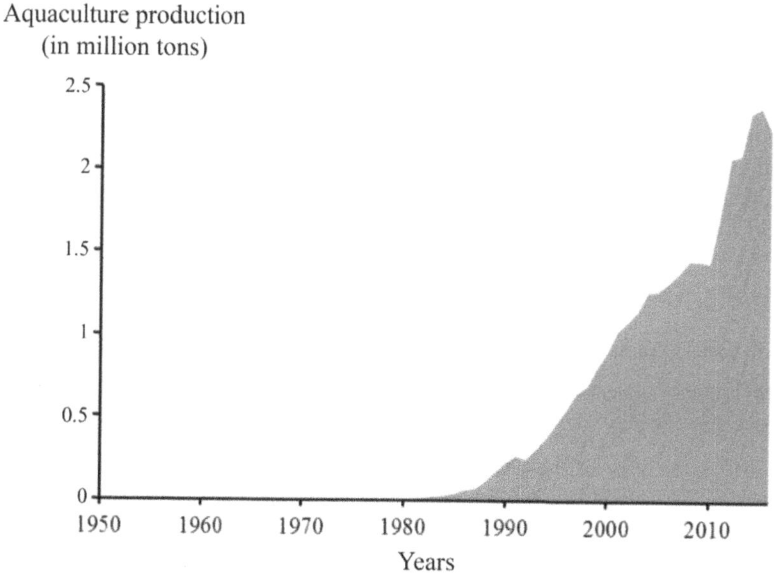

Figure 6.4. *Global Atlantic salmon production from 1950 to 2016 (source: F. Teletchea)*

In just a few decades, Atlantic salmon has gone from being a luxury product, consumed only at specific times of the year (notably Christmas), to an everyday consumer item. In 2014, it ranked ninth worldwide among aquatic products, with over 2.3 million tons sold. The success of salmon farming in Norway and Chile is mainly due to the presence of many suitable production sites, dynamic research and industry, as well as government support. In addition, the salmon has a number of features (high growth in cages, flesh highly appreciated by consumers, etc.) that contribute to its farming success. At the same time, catches by fisheries have fallen sharply, from 118,000 tons in 1980 to 73,000 tons in 1990, and to less than 40,000 tons today. In conclusion, almost 100% of the Atlantic salmon consumed worldwide today comes from farming (Teletchea 2016).

6.5. Conclusion

Domestication is an ongoing process that adapts fish to both captive conditions and humans. Domestication only began a few decades (or even years) ago for most farmed fish species, so probably less than one-third could be considered domesticated. Many traits, including growth, were modified during domestication. New breeding programs will need to balance market and non-market values (fish welfare), while retaining sufficient genetic variability to ensure that fish are productive and robust enough to cope with environmental changes (global warming, ocean acidification and declining dissolved oxygen in water). The sustainable future of aquaculture will first depend on the continuous improvement of fish species that have already been domesticated, and, second, on our willingness and ability to diversify farmed fish species, preferably indigenous ones, to promote more resilient and balanced aquaculture production worldwide.

6.6. References

Balon, E. (2004). About the oldest domesticates among fishes. *J. Fish Biol.*, 65(S1), 1–27.

Bilio, M. (2008). Controlled reproduction and domestication in aquaculture. Report, European Aquaculture Society, Oostende.

FAO (2019). The state of the world's aquatic genetic resources for food and agriculture. Report, FAO Commission on Genetic Resources for Food and Agriculture Assessments, Rome.

Fontaine, P. and Teletchea, F. (2019). Domestication of Eurasian perch. In *Animal Domestication*, Teletchea, F. (ed.). IntechOpen, London.

Gjedrem, T. (2012). Genetic improvement for the development of efficient global aquaculture: A personal opinion review. *Aquaculture*, 12(22), 344–349.

Gjedrem, T. and Baranski, M. (2009). *Selective Breeding in Aquaculture: An Introduction. Reviews: Methods and Technologies in Fish Biology and Fisheries*. Springer, Berlin.

Gjedrem, T., Robinson, N., Rye, M. (2012). The importance of selective breeding in aquaculture to meet future demands for animal protein: A review. *Aquaculture*, 117–129, 350–353.

Glover, K.A., Solberg, M., McGinnity, P., Hindar, K., Verspoor, E., Coulson, M.W., Hansen, M.M., Araki, H., Skaala, Ø., Svåsand, T. (2017). Half a century of genetic interaction between farmed and wild Atlantic salmon: Status of knowledge and unanswered questions. *Fish Fish.*, 18, 890–927.

Glover, K.A., Wennevik, V., Hindar, K., Skaala, O., Fiske, P. (2020). The future looks like the past: Introgression of domesticated Atlantic salmon escapees in a risk assessment framework. *Fish Fish.*, 21, 1077–1091.

Houston, R.D., Bean, T.P., Macqueen, D.J., Gundappa, M.K., Jin, Y.H., Jenkins, T.L., Selly, S.L.C., Martin, S.A.M., Stevens, J.R., Santos, E.M. et al. (2020). Harnessing genomics to fast-track genetic improvement in aquaculture. *Nat. Rev. Genet.*, 21, 389–409.

Janssen, K., Chavanne, H., Berentsen, P., Komen, H. (2017). Impact of selective breeding on European aquaculture. *Aquaculture*, 472(1), 8–16.

Nakajima, T., Hudson, M.J., Uchiyama, J., Makibayashi, K., Zhang, J. (2019). Common carp aquaculture in Neolithic China dates back 8,000 years. *Nat. Ecol. Evol.*, 3, 1415–1418.

Teletchea, F. (2012). Qu'est-ce qu'un poisson domestique ? Implications pour le développement futur de l'aquaculture. *Ethnozootechnie*, 90, 7–12.

Teletchea, F. (2016). *De la pêche à l'aquaculture. Demain, quels poissons dans nos assiettes ?* Belin, Paris.

Teletchea, F. (2019). Fish domestication in aquaculture: Reassessment and emerging questions. *Cybium*, 43, 7–15.

Teletchea, F. (2021). Fish domestication in aquaculture: 10 unanswered questions. *Anim. Front.*, 11(3), 87–91.

Vandeputte, M. (2009). Méthodes de gestion et d'amélioration génétique des populations d'élevage. In *La truite arc-en-ciel : de la biologie à l'élevage*, Jalabert, B. and Fostier, A. (eds). Quae Editions, Versailles.

7

The Domestication of Yeast

Jean-Luc LEGRAS, Thibault NIDELET, Virginie GALEOTE and Delphine SICARD

SPO, INRAE, Montpellier, France

Yeasts are single-celled fungi that are involved in the production of many of the fermented foods and beverages that make up our cultural heritage. Fermented products were probably already consumed by humans when they were hunter-gatherers. Their consumption then developed in the early ***Holocene era***, with the emergence of agriculture and the domestication of plants and animals. Their production involves yeasts, filamentous fungi and bacteria, which were first used subconsciously, then consciously by humans. This chapter focuses on the domestication of yeast. After briefly outlining the origins of fermented products, we will describe the diversity of yeast species encountered, then present what we know about the evolutionary history of domesticated yeasts. Lastly, we will detail the genetic events that helped yeasts to adapt to their domesticated environment, as well as the traits that meet the needs of humans.

7.1. The history of fermented products and the domestication of microorganisms

A wide variety of fermented foods and beverages are consumed around the world. Whether based on plant or animal products, most have been produced since ancient times: wine, beer, sake, soy sauce, bread, cheese, sausage, fruit kefir, kombucha, etc., the list is long and not exhaustive, as a result of culinary creativity.

It is thought that these products were originally developed for their ability to be preserved. It is also possible that alcoholic beverages had medicinal, social and spiritual functions from the very start. Fermented products were then developed to improve the organoleptic, nutritional quality and diversity of foods and beverages.

It is difficult to date the appearance of the first fermented products. In Asia, archaeologists have found evidence of fermented beverage production from rice, honey and fruit dating back to 7,000 BCE in the Neolithic village of Jiahu, China. Beer made from sprouted millet has been identified in containers from three sites of the Yangshao culture in northern China (ca. 3,700–2,700 BCE). Low-alcohol beverages made from millet, *Triticeae* grasses, Job's tears, rice, beans, *Trichosanthes cucumerina* root, ginger, yam, lily, etc., could have also been produced at this time in China (Liu et al. 2019).

In Mesopotamia and around the Mediterranean, the first evidence of fermented products was found in the Raquefet cave in Israel, revealing beer production from wild wheat and barley 13,000 years ago. Evidence of wine-making is based on the presence of tartaric acid in an ancient jar, dating from 5,400 to 5,000 BCE at the Neolithic site of Tepe, Mesopotamia, and the remains of grape juice extraction, dating from 4,000 BCE at the Neolithic site of Dikili Tash, Greece. Those of yeast bread were found in Ancient Egypt at the same time.

Since the Neolithic period, different fermentation practices, associated with different levels of human intervention, have been and continue to be used. Fermentation can be spontaneous, taking place all at once in the plant or animal matrix. It can also be based on the preparation of a ferment that is continuously maintained over months, and even years, by the recurrent addition of plant or animal resources, a process known as backslopping. Lastly, it can be done using an industrial ferment, composed of one or more selected microbial strains. These different practices are associated with different levels of human intervention in microorganisms, ranging from light management of a microbial community to the controlled reproduction of a few strains (Figure 7.1). The domestication of microorganisms, defined as the process of selection for adaptation to agroecological niches and human preferences, is similar to the domestication of animals and plants. Furthermore, beyond selection, humans influence the evolution of microorganisms by controlling or influencing their dispersal, or by indirectly modifying their mutation rate. There is therefore a continuum of practices and relationships between humans and microorganisms, which have influenced and continue to influence the paths, objectives and speed of domestication.

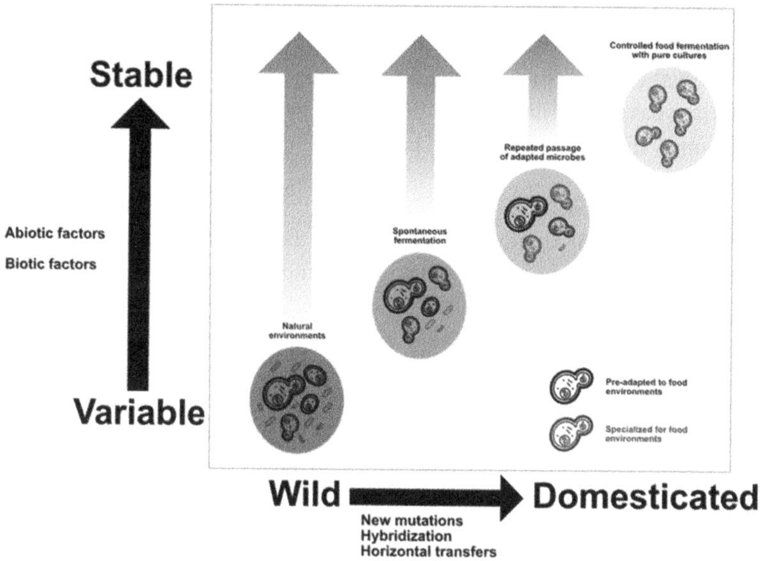

Figure 7.1. *Microorganism domestication continuum (adapted from Gibbons and Rinker (2015)). For a color version of this figure, see www.iste.co.uk/pelletier/genetics.zip*

COMMENT ON FIGURE 7.1.– *This model shows the domestication of a yeast species. The Y-axis represents the variability of fermentation practices and the associated biotic and abiotic environments that the yeast species may encounter, while the X-axis reflects the progression from wild to domesticated. The natural environment is generally more variable than the food environment, and the range of relative variability is represented by light blue arrows at each stage. At the spontaneous fermentation stage (e.g. kimchi and cocoa), microbes that are pre-adapted to the substrate and fermentation environment dominate the microbial community. Repeated long-term passages of pre-adapted microbes (in other words, either through persistence in the food processing environment or intentional backslopping) into the more controlled food environment promote adaptation and specialization (blue yeast, in this case), which can be achieved through a variety of mechanisms, including new mutations, hybridization and horizontal gene transfer. Lastly, in special cases, such as industrial food production, abiotic and biotic factors are carefully regulated, and pure cultures of the domesticated microbe are used to guarantee the quality, homogeneity and safety of the food product.*

7.2. Yeast diversity and the evolutionary origins of fermentation

The microbial communities involved in beverage and food production are made up of bacteria, yeasts and filamentous fungi. The proportion of these three types of microorganisms varies according to the product.

Here, we have chosen to focus on yeasts. A yeast can be defined as a single-celled fungus that divides by budding or fission and, while it reproduces sexually, does not dispose of its spores inside fruiting bodies (Kurtzman et al. 2011).

Yeasts have a ***polyphyletic*** origin and are found in the *Ascomycota* and *Basidiomycota* phylum. Over 1,350 species of yeast have been listed and validated. Some of them are capable of alcoholic fermentation, that is to say, capable of transforming carbohydrates, usually glucose, into ethanol and carbon dioxide. This fermentative metabolism has been used by mankind to transform agricultural raw materials, first empirically, then rationally.

Yeasts are capable of carrying out alcoholic fermentation in the absence of oxygen, but also in the presence of oxygen when sugar is in sufficient supply, a capability known as the "Crabtree effect". "Crabtree-positive" yeasts convert carbohydrates into ethanol, and when sugar is in short supply, they are able to switch from fermentative metabolism[1] to respiratory metabolism[2], where the ethanol is consumed. This change is known as the *diauxic shift* and may be associated with better ethanol tolerance and a competitive advantage for yeast in alcoholic products.

Several evolutionary genetic studies have traced the history of the acquired ability to ferment in the absence and presence of oxygen in the *Saccharomycetaceae*[3]. It appears to have arisen as a result of several independent events that gradually reshaped the metabolism. These events include the loss of respiratory complex I[4], a ***horizontal transfer*** of the *URA1* gene involved in

1 All the chemical reactions that transform sugars into alcohol, carbon dioxide and/or organic acid. It occurs in the absence of oxygen, or in some yeasts in the presence of O_2, when there is a high quantity of sugar.

2 All the chemical reactions that transform sugars in the presence of oxygen and produce large quantities of energy.

3 The *Saccharomycetaceae* are a family of fungi in the order Saccharomycetales, also known as the "budding yeasts" order, whose best-known species is *Saccharomyces cerevisiae*, the baker's and brewer's yeast.

4 NADH dehydrogenase, the first enzyme in cellular respiration.

pyrimidine base biosynthesis, the duplication of glycolysis[5] and glucose transporter genes, the ability to grow without ***mitochondrial DNA***, and the reorganization of the ***transcription*** regulatory network (Hagman and Piškur 2015). These events are thought to have occurred in gradual succession over the last 150 million years, culminating in the development of the ability to ferment in the absence of oxygen in five yeast genera: *Tetrapisispora*, *Vanderwaltozyma*, *Nakazeomyces*, *Kazachstania* and *Saccharomyces*, where it is most highly developed.

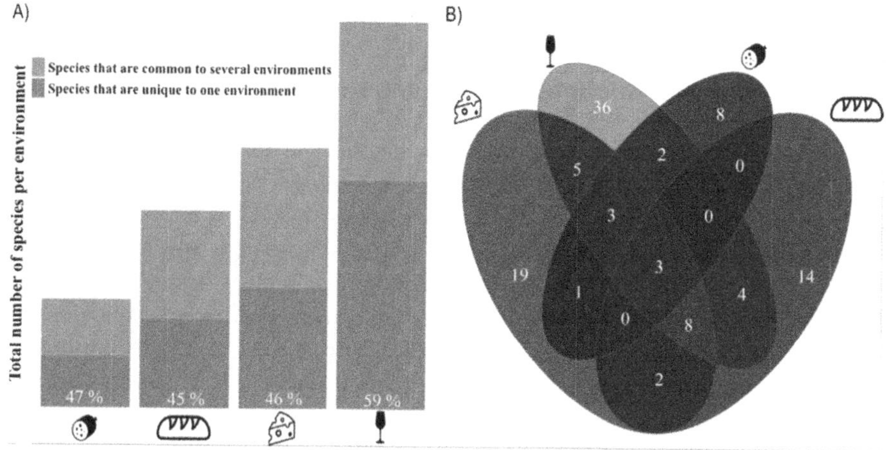

Figure 7.2. *Diversity and specialization of yeast species in four fermented foods (source: J.-L. Legras). For a color version of this figure, see www.iste.co.uk/pelletier/genetics.zip*

COMMENT ON FIGURE 7.2.– *(a) Number of different and unique yeast species for four fermented foods (meat, bread, cheese and wine). The height of the bars represents the total number of species. The blue part of the bars represents species that are unique to that environment; (b) Venn diagram of yeast species distribution for four fermented foods. The color is proportional to the number of species in each zone, which is also indicated by a white number.*

A wide diversity of yeasts can be identified in the microbial communities involved in food and beverage production. Analysis of 85 articles describing yeast diversity in five products (chocolate, wine, bread, cheese and beer) and associated environments (must, milk and insects) revealed that over 320 yeast species had been

5 A set of metabolic reactions producing energy, by converting glucose into two molecules of pyruvate.

identified in these products (Figure 7.2). Some species are ubiquitous and found in all or many fermented products, such as *Saccharomyces cerevisiae*, *Torulaspora delbrueckii*, *Kluyveromyces marxianus*, *Pichia fermentans*, *Debaryomyces hansenii* and *Brettanomyces bruxellensis*. Others are more specific, such as *Kazachstania bulderi*, which can only be identified in cereal products, or *Starmerella bacillaris* in grape must. Phylogenetically related species do not preferentially occur in one type of product, suggesting that there is no obvious link between fungal ***microbiota*** composition and ***phylogeny***.

7.3. Population structure of yeast isolated from anthropic niches

Numerous yeast isolation campaigns in natural and man-made environments have been carried out since the end of the 19th century. They have revealed that many yeast species are found both in natural habitats, such as water, soil or forests, and in habitats such as fermented products. For a long time, it was impossible to analyze their genetic diversity. The rise of molecular biology techniques, and in particular ***PCR amplification***, has led to the development of different generations of molecular methods, providing more numerous, more specific and therefore higher quality markers. The latest generations of genetic markers (AFLP[6], ***microsatellites*** and ***MLST***) and genome sequencing have made it possible to study the population structure of several yeast species.

7.3.1. *No genetic structure*

Several species found in grape musts do not show genetic structuring related to their anthropic niche (reviewed in Albertin et al. (2019, pp. 117–163)). This is the case for *Hanseniaspora uvarum*, the most abundant species in musts, at the beginning of grape fermentation. Although these populations genetically group according to geographical location and time, wine strains do not appear to be genetically specific compared to those from natural environments. Similarly, analysis of strains of *Starmerella bacillaris*, the second most abundant species in musts, revealed no clear structuring. Ultimately, yeasts of the *Saccharomyces uvarum* species show a strong geographical structure (Holarctic/South America and Australia), but genome analysis of 32 strains isolated from wine or cider from northern regions and natural environments shows that wine and cider strains are not genetically different.

6 AFLP (Amplified Fragment Length Polymorphism): size ***polymorphism*** of restriction enzyme-digested DNA fragments, amplified by PCR.

7.3.2. *Genetic structure associated with anthropic niches*

7.3.2.1. *Monophyletic origin*

While populations of the two most abundant species in grape must do not seem to differentiate according to their habitat of origin, other species, found on grape berries, show genetically specific populations. Wine strains of *S. cerevisiae*, isolated worldwide, are genetically combined into a single group, suggesting a monophyletic origin (Legras et al. 2018; Peter et al. 2018). The same is true of the *Torulaspora delbrueckii* and *Lachancea thermotolerans* species, which have a positive effect on wine traits, and for the wine-altering yeast *Brettanomyces bruxellensis*, whose wine populations are different from those found in beer or Kombucha (Albertin et al. 2019, pp. 117–163). It also appears that by developing winemaking, humans have created a unique environment that has profoundly altered the diversity of certain yeast species, allowing the emergence of specific genotypes. This environment has been propagated worldwide with the spread of vine cultivation, and population genetics data strongly suggest a transfer of these microbial communities with grapevines.

There are also several examples among yeasts isolated from milk and cheese. The yeasts *Kluyveromyces lactis var lactis* and *Kluyveromyces marxianus* have acquired the ability to assimilate lactose, thanks to the acquisition of a lactose transporter that differentiates them from those of natural habitats (Ortiz-Merino et al. 2018; Varela et al. 2018). Cheese-specific genetic groups have also been identified for the *S. cerevisiae*, *Geotrichum candidum* and *T. delbrueckii* yeast species (Legras et al. 2018).

7.3.2.2. *Polyphyletic origin*

The evolutionary history of *S. cerevisiae* populations from bread, chocolate, coffee and beer seems more complex than that of wine. In the case of bread, *S. cerevisiae* strains can be divided into two genetically distinct groups, one containing a majority of industrial strains of ***tetraploid*** baker's yeast and the other containing strains isolated from bread-making sourdough, which are most often ***diploid***. The commercial strains show traces of beer strain introgression, while the sourdough strains show traces of introgression from Asian fermentation strains, such as sake. Baker's yeast populations have thus followed two evolutionary trajectories, depending on whether they have been used in industrial baking or artisanal baking (Bigey et al. 2021).

The evolutionary history of *S. cerevisiae* beer strains is also more complex than that of wine populations. Today's industrial yeasts can be divided into two major lineages that are genetically and phenotypically separate from wild strains

(Gallone et al. 2018). They are thought to have derived from just a few ancestors and to have diverged as a result of domestication processes. One of the lineages is itself separated into three sub-groups, respectively, bringing together strains from England, Belgium/Germany and the United States, suggesting geographical isolation.

Lastly, *S. cerevisiae* strains isolated from coffee drupe and cocoa bean fermentations are genetically different, clearly showing distinct origins depending on the fruit's continent of origin (Ludlow et al. 2016). For example, coffee drupe populations are clearly separate depending on whether they come from South America or Africa. As with beer, there seems to be a certain geographical isolation between domesticated populations.

Overall, these results show that not all yeast populations isolated from anthropic niches have diverged, nor are they all potentially domesticated. When they are, their evolutionary history either reflects the expansion of the crops that make up the fermented matrix (e.g. wine), the fermentation practices (e.g. bread) or geographical isolation (e.g. coffee).

7.4. Genetic basis for the evolutionary history of domesticated populations

7.4.1. *Hybridization and ploidy*

Polyploids (3-5n) are frequently encountered in yeast, particularly in human-related niches. This may be attributed to the advantage given by genome duplication (ability to grow improved by 3.6% in certain environments; Fisher et al. 2018). Polyploid formation can result from hybridization between strains of different species (allopolyploid), of the same species (autopolyploid), or from chromosome duplication without cell division (endopolyploid). Autopolyploids and allopolyploids are frequently identified in yeast populations of the genus *Saccharomyces*, destined for food or alcoholic beverages. For example, commercial bread-making strains are most often tetraploid (Bigey et al. 2021), and ***allele*** segregation at ***meiosis*** suggests that these polyploids are autotetraploids resulting from intraspecific hybridization between *S. cerevisiae* strains.

Many *Saccharomyces* strains used in winemaking have also been shown to be polyploid. This is the case, for example, with *S. cerevisiae* strains that vary in ploidy for Spanish sherry-type wines.

Beyond the examples of autotetraploids, the genus *Saccharomyces* is illustrated by a large number of allopolyploids or diploids with one or more interspecific hybridization events in their evolutionary history. The genus *Saccharomyces* currently comprises eight species (*S. cerevisiae*, *S. eubayanus*, *S. uvarum*, *S. paradoxus*, *S. mikatae*, *S. jurei*, *S. arboricola* and *S. kudriavzevii*). All were found in natural environments, in soil or tree bark. The absence of prezygotic barriers[7] between these species would have enabled hybrids to form in both natural and domesticated habitats. Humans, through directed selection, have also contributed to the creation of new hybrids.

A well-known example of an interspecific hybrid in the genus *Saccharomyces* is *Saccharomyces pastorianus*, which is widely used in brewing to produce lager beer. Progenitors of *S. pastorianus* have been identified as *S. cerevisiae* and *Saccharomyces eubayanus*. A second example is *Saccharomyces bayanus*, also used to make beer, which comes from a triple hybridization between *S. eubayanus*, *S. uvarum* and *S. cerevisiae*. A summary of interspecific hybridizations in the genus *Saccharomyces* can be found in the review (Alsammar and Delneri 2020), which Figure 7.3 is taken from.

Ploidy variations and examples of interspecific hybridization also exist in other yeast genera. For example, populations of *Brettanomyces bruxellensis*, a yeast that contributes to the specificity of certain beers, but is unfavorable for wine flavors, present different levels of ploidy with diploids, triploids or tetraploids.

These ploidy variations appear to be due, at least in part, to interspecific hybridization events with currently unknown species (Eberlein et al. 2021). Similarly, diploid, triploid and tetraploid interspecific hybrids have been detected between *H. opuntiae* and *H. pseudoguilliermondii*, two other species found in wine (Saubin et al. 2019).

7.4.2. *Horizontal transfers and introgressions*

Genomic analyses of *S. cerevisiae* populations have revealed the presence of regions from other microbial species (Legras et al. 2018). Wine yeasts are a remarkable example of this through the presence, in the genome, of a majority of strains, of two regions, called B and C, derived from two distant species: *Zygosaccharomyces bailii* and *Torulaspora microellipsoides*, respectively.

7 Which precede embryo formation.

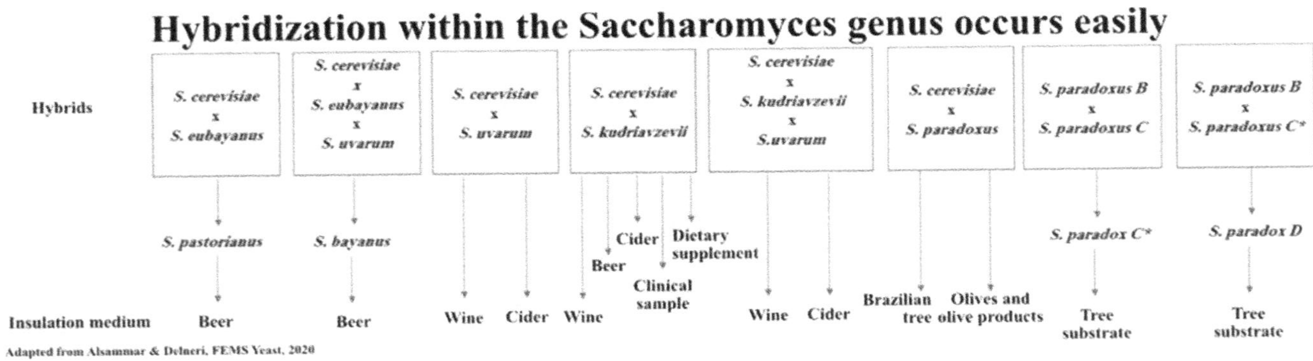

Figure 7.3. *Common Saccharomyces hybrids with their isolation medium (adapted from Alsammar and Delneri (2020)). Saccharomyces species can hybridize to form hybrids with industrial significance. Most known hybrids are associated with domestication activities and only a few are isolated in non-fermentation environments. For a color version of this figure, see www.iste.co.uk/pelletier/genetics.zip*

These regions, 17 kb and 65 kb in size, contain 5 and 19 genes, respectively, potentially encoding proteins that are mainly involved in important metabolic functions during winemaking, such as carbon and nitrogen metabolism, cell transport and stress response. While the contribution of the B-region genes has not yet been established, the C-region includes several genes, including a high-affinity fructose transporter (*FSY1*), a function that was previously absent in *S. cerevisiae*, one that may give it an advantage at the end of fermentation by assimilating fructose, which under these conditions becomes the main residual sugar, and the small peptide transporters *FOT1* and *FOT2*. These two genes give a significant competitive advantage to yeasts that possess them during oenological fermentation. In addition, these genes modify the metabolite profile, resulting in a marked reduction in acetic acid and an increase in certain aromatic compounds produced during fermentation. These organoleptic traits, which probably improve the aromatic profile of wines, suggest a possible influence of domestication in the acquisition of *FOT* genes by *S. cerevisiae* wine strains. Humans, in their desire to obtain pleasant aromatic wines and improve the fermentation process, may have indirectly selected yeast inoculants with strains containing these genes.

Another outstanding example is a 6.8 kb genomic region, known as T, found exclusively in cheese strains of *S. cerevisiae*. This region contains genes homologous to *S. cerevisiae GAL7-GAL10-GAL1* (46% nucleic identity) involved in galactose metabolism. Derived from a species close to *Saccharomyces*, but unknown to date, it has replaced the GAL cluster of *S. cerevisiae*. While cheese strains, unlike wine strains, are not capable of complete oenological fermentation, they have a better ability to metabolize galactose (sugar derived from the hydrolysis of milk lactose), with both a reduced latency phase and a higher growth rate than all strains of other origins. *Kluyveromyces lactis var lactis* yeasts similarly acquired the ability to metabolize lactose by integrating a gene into their genome that encodes the Lac12 transporter from a *Kluyveromyces marxianus* population (Varela et al. 2018).

7.4.3. *Chromosomal rearrangement*

Among yeasts, the most remarkable chromosomal rearrangements are probably those observed in the wine yeast *S. cerevisiae*. The systematic use of sulfite in winemaking has promoted the emergence of resistant mutants. A first translocation between chromosomes 7 and 16, located just before the *SSU1* gene encoding a transporter expelling sulfite from the cell, was characterized as early as 2002 (Pérez-Ortín et al. 2002). It was then shown that this translocation allows the ***promoter*** of this gene to be replaced by a new region, containing repeats. Surprisingly, the number of repeats in this new promoter is correlated with gene expression and sulfite resistance. However, since this first example, two other types of translocation at this site have been characterized, all leading to overexpression of the *SSU1* gene. Lastly, similar translocations have been identified in *Saccharomyces uvarum* yeasts isolated from wine (Macías et al. 2021). Indeed, when fermenting wine or cider at low temperatures (<10°C), this cold-tolerant species, a cousin of *Saccharomyces cerevisiae*, takes its place.

7.4.4. *Duplication*

Genome, chromosome or gene duplication events lead to the appearance of new genetic sources of adaptation. These events have repeatedly played a part in the evolution of domesticated *S. cerevisiae* populations.

The ancestor of the *S. cerevisiae* species duplicated its genome around 100 million years ago, probably as a result of interspecific hybridization. Duplication of the genome would have restored fertility to this hybrid (Marcet-Houben and Gabaldón 2015). Some of the duplicated genes remained. At present, the *S. cerevisiae* genome contains around 5,000 genes, 1,100 of which are duplicated. Half of these come from ancestral genome duplication, with conserved copies forming ohnolog pairs[8].

The fraction of genes regulated under stress conditions is significantly higher among ohnologs than among single genes in *S. cerevisiae*, suggesting that ancestral duplication of the *S. cerevisiae* genome is partly responsible for population adaptation to environmental stresses. Fermented products are generally environments with extreme characteristics that generate stress: osmotic pressure, ethanol content, concentration of copper sulfate used as a phytosanitary agent, salt content, etc. Indeed, several domesticated populations have maintained duplicated genes that are known to be involved in tolerance to these stresses. This is the case, for example, for the *CUP1* gene, which encodes a protein that chelates copper in yeast, and which is duplicated in tandem in wine and sake strains. This gene varies from one to 79 copies in wine strains, and these copy variations are partly associated with variations in copper tolerance (Crosato et al. 2020).

Several genes involved in carbon metabolism are also duplicated in domesticated populations. The *MAL* genes involved in maltose (glucose dimer) consumption are often duplicated in *S. cerevisiae* beer or bread-making sourdough populations, where maltose is an important resource (Gallone et al. 2016; Bigey et al. 2021), whereas

8 Two genes that perform different functions and have divergently evolved from the same ancestral gene after duplication of the entire genome.

they are not in wine populations, where maltose is not a carbon source. The *GAL2* gene, which encodes a permease that enables galactose import, is duplicated in cheese strains (Legras et al. 2018). The fact that these duplications are only maintained in strains that are isolated from the environment containing the sugar in question suggests that they are adaptive.

7.4.5. *SNP: selection signature a priori*

The alleles of certain genes can also show a high number of differences between populations, reflecting a specialization of these alleles according to the ecological niche. For example, the *S. cerevisiae* cheese yeast population has an allele of the *GAL2* high-affinity transporter that is very different from alleles detected in other populations. The transporter encoded by the *GAL2* gene in cheese yeasts seems less specialized than that of strains isolated from other environments, as it is capable of transporting glucose in addition to galactose. The same is true of flor yeast strains, which develop on the wine surface after alcoholic fermentation (and sugar depletion), forming a biofilm and producing an oxidative metabolism based on ethanol consumption. For these yeasts, the ability to produce a film is associated with the Flo11 protein. These yeasts possess an *RGA2* gene allele that is very different from wine yeasts, which favors biofilm production (Coi et al. 2017). Similarly, a truncated form of the Sfl1 transcription factor, which regulates *FLO11*, was observed in some yeasts. Lastly, this film yeast population exhibits divalent metal ion transporters (Zn^{2+}, Mn^{2+}, Mg^{2+}) that are different from those of wine yeasts, with the high-affinity zinc transporter Zrt1 specifically under positive selective pressure in this environment (Coi et al. 2017).

7.5. Conclusion

Recent genomic studies have revealed that yeast domestication for adaptation to different ecological niches results from a combination of structural variations[9] and point mutations that only affect one or a few nucleotides (Figure 7.4).

S. cerevisiae brewer's yeast has mutations in several genes that enable it to acquire specific technological properties.

9 Modification of a large part of the genome (amplification of the number of gene copies, elimination or insertion of chromosome fragments, etc.).

These yeasts combine the amplification of several genes involved in maltose hydrolysis, with the presence of a particular allele of the *MAL11* genes (*ATG1*) giving them the ability to penetrate maltotriose (glucose trimer).

Furthermore, most of these brewer's yeasts have lost the ability to produce vinyl guaiacol, an aroma that is rejected by consumers, due to loss-of-function mutations in the *PAD1* and *FDC1* genes responsible for synthetizing this compound.

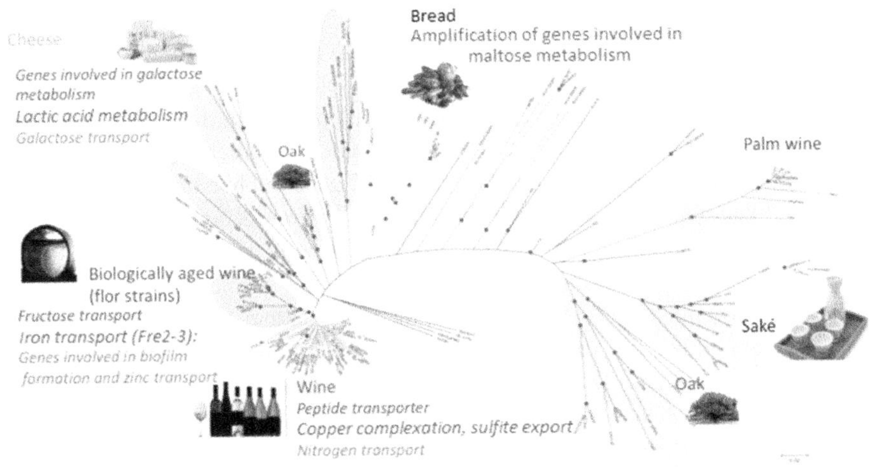

Figure 7.4. *Genomics of domestication in Saccharomyces cerevisiae. Functions of horizontally transferred genes (red), translocation or amplification (green), or mutation (blue) involved in adaptation to different niches (source: modified from Legras et al. (2018))*

For cheese and milk, strains of the *Kluyveromyces lactis var lactis*, *K. marxianus* and *S. cerevisiae* species carry different mutations that enable them to better transport and metabolize lactose or galactose.

In the same way, the ability of *S. cerevisiae* yeasts to build the biofilm that forms the film needed to age sherry and Jura wines is linked to the specificities acquired by flor yeasts. The number of mutations in the *FLO11* gene, its promoter and regulators is remarkable.

However, beyond these mutations in a few target genes, comparison of wine and flor yeast genomes reveals clear differentiations in multiple regions, which suggests their involvement in adaptation, as in the case of divalent metal ion transporters, although the mechanism remains unknown.

Lastly, wine yeasts stand out for their ability to complete the fermentation of grape musts, which is better than that of yeasts of other origins. Analyses of their genomes also show that adaptation to this ecological niche is the result of several modifications. The integration of genes from distant species by horizontal transfer to improve peptide assimilation capacity and ultimately numerous signs of selection of certain alleles seem to point to amino acid transporters or sterol transport and many other targets. Translocations leading to the overexpression of the *SSU1* gene, or the amplification of *CUP1* genes, correspond, in turn, to the recent adaptation to chemical additives used to control the sanitary aspects of vine cultivation and fermentation.

These examples of domestication illustrate the remarkable adaptability of yeasts, enabling them to colonize new habitats created by humans.

7.6. References

Albertin, W., Chasseriaud, L., Comte, G., Panfili, A., Delcamp, A., Salin, F., Marullo, P., Bely, M. (2014). Winemaking and bioprocesses strongly shaped the genetic diversity of the ubiquitous yeast *Torulaspora delbrueckii*. *PLoS ONE*, 9(4). doi: 10.1371/journal.pone.0094246.

Albertin, W., Masneuf-Pomarede, I., Galeote, V., Legras, J.L. (2019). New insights into wine yeast diversities. In *Yeasts in the Production of Wine*, Romano, P., Ciani, M., Fleet, G. (eds). Springer, Berlin.

Alsammar, H. and Delneri, D. (2020). An update on the diversity, ecology and biogeography of the Saccharomyces genus. *FEMS Yeast Research*, 20(3), foaa013. doi: 10.1093/femsyr/foaa013.

Bigey, F., Segond, D., Friedrich, A., Guezenec, S., Bourgais, A., Huyghe, L., Agier, N., Nidelet, T., Sicard, D. (2021). Evidence for two main domestication trajectories in *Saccharomyces cerevisiae* linked to distinct bread-making processes. *Current Biology*, 31(4), 722–732.e5. doi: 10.1016/j.cub.2020.11.016.

Coi, A.L., Bigey, F., Mallet, S., Marsit, S., Zara, G., Gladieux, P., Galeote, V., Budroni, M., Dequin, S., Legras, J.L. (2017). Genomic signatures of adaptation to wine biological ageing conditions in biofilm-forming flor yeasts. *Molecular Ecology*, 26(7), 2150–2166. doi: 10.1111/mec.14053.

Crosato, G., Nadai, C., Carlot, M., Garavaglia, J., Ziegler, D.R., Rossi, R.C., De Castilhos, J., Campanaro, S., Treu, L., Giacomini, A., Corich, V. (2020). The impact of CUP1 gene copy-number and XVI-VIII/XV-XVI translocations on copper and sulfite tolerance in vineyard *Saccharomyces cerevisiae* strain populations. *FEMS Yeast Research*, 20(4), foaa028. doi: 10.1093/femsyr/foaa028.

Eberlein, C., Abou Saada, O., Friedrich, A., Albertin, W., Schacherer, J. (2021). Different trajectories of polyploidization shape the genomic landscape of the *Brettanomyces Bruxellensis* yeast species. *Genome Research*, 31(12), 2316–2326.

Fisher, K.J., Buskirk, S.W., Vignogna, R.C., Marad, D.A., Lang, G.I. (2018). Adaptive genome duplication affects patterns of molecular evolution in *Saccharomyces cerevisiae. PLOS Genetics*, 14(5), e1007396. doi: 10.1371/journal.pgen.1007396.

Gallone, B., Mertens, S., Gordon, J.L., Maere, S., Verstrepen, K.J., Steensels, J. (2018). Origins, evolution, domestication and diversity of Saccharomyces beer yeasts. *Curr Opin Biotechnol*, 49, 148–155. doi: 10.1016/j.copbio.2017.08.005.

Gibbons, J.G. and Rinker, D.C. (2015). The genomics of microbial domestication in the fermented food environment. *Curr Opin Genet Dev*, 35, 1–8. doi: 10.1016/j.gde.2015.07.003.

Hagman, A. and Piškur, J. (2015). A study on the fundamental mechanism and the evolutionary driving forces behind aerobic fermentation in yeast. *PLoS ONE*, 10(1), 1–24. doi: 10.1371/journal.pone.0116942.

Kurtzman, C.P., Fell, J.W., Boekhout, T. (2011). *The Yeasts: A Taxonomic Study*. Elsevier, Amsterdam.

Legras, J.-L., Galeote, V., Bigey, F., Camarasa, C., Marsit, S., Nidelet, T., Sanchez, I., Couloux, A., Guy, J., Franco-Duarte, R. et al. (2018). Adaptation of *S. cerevisiae* to fermented food environments reveals remarkable genome plasticity and the footprints of domestication. *Molecular Biology and Evolution*, 35(7), 1712–1727. doi: 10.1093/molbev/msy066.

Liu, L., Wang, J., Levin, M.J., Sinnott-Armstrong, N., Zhao, H., Zhao, Y., Shao, J., Di, N., Zhang, T. (2019). The origins of specialized pottery and diverse alcohol fermentation techniques in early Neolithic China. *Proceedings of the National Academy of Sciences*, 116(26), 12767–12774. doi: 10.1073/pnas.1902668116.

Ludlow, C.L., Cromie, G.A., Garmendia-Torres, C., Sirr, A., Hays, M., Field, C., Jeffery, E.W., Fay, J.C., Dudley, A.M. (2016). Independent origins of yeast associated with coffee and cacao fermentation. *Current Biology*, 26(7), 965–971. doi: 10.1016/j.cub.2016.02.012.

Macías, L.G., Flores, M.G., Adam, A.C., Rodríguez, M.E., Querol, A., Barrio, E., Lopes, C.A., Pérez-Torrado, R. (2021). Convergent adaptation of *Saccharomyces uvarum* to sulfite, an antimicrobial preservative widely used in human-driven fermentations. *PLOS Genetics*, 17(11), e1009872. doi: 10.1371/journal.pgen.1009872.

Marcet-Houben, M. and Gabaldón, T. (2015). Beyond the whole-genome duplication: Phylogenetic evidence for an ancient interspecies hybridization in the baker's yeast lineage. *PLoS Biology*, 13(8), 1–26. doi: 10.1371/journal.pbio.1002220.

Ortiz-Merino, R.A., Varela, J.A., Coughlan, A.Y., Hoshida, H., da Silveira, W.B., Wilde, C., Kuijpers, N.G.A., Geertman, J.M., Wolfe, K.H., Morrissey, J.P. (2018). Ploidy variation in *Kluyveromyces marxianus* separates dairy and non-dairy isolates. *Frontiers in Genetics*, 9, 94. doi: 10.3389/fgene.2018.00094.

Pérez-Ortín, J.E., Querol, A., Puig, S., Barrio, E. (2002). Molecular characterization of a chromosomal rearrangement involved in the adaptive evolution of yeast strains. *Genome Research*, 12(10), 1533–1539. doi: 10.1101/gr.436602.

Peter, J., De Chiara, M., Friedrich, A., Yue, J.-X., Pflieger, D., Bergström, A., Sigwalt, A., Barre, B., Freel, K., Llored, A. et al. (2018). Genome evolution across 1,011 *Saccharomyces cerevisiae* isolates. *Nature*, 556(7701), 339–344. doi: 10.1038/s41586-018-0030-5.

Saubin, M., Devillers, H., Proust, L., Brier, C., Grondin, C., Pradal, M., Legras, J.L., Neuvéglise, C. (2020). Investigation of genetic relationships between Hanseniaspora species found in grape musts revealed interspecific hybrids with dynamic genome structures. *Frontiers in Microbiology*, 10. doi: 10.3389/fmicb.2019.02960.

Varela, J.A., Puricelli, M., Ortiz-Merino, R.A., Giacomobono, R., Braun-Galleani, S., Wolfe, K.H., Morrissey, J.P. (2019). Origin of lactose fermentation in *Kluyveromyces lactis* by interspecies transfer of a neo-functionalized gene cluster during domestication. *Current Biology*, 29(24), 4284–4290.e2. doi: 10.1016/j.cub.2019.10.044.

8

The Domestication of *Oenococcus oeni*: A Bacterium Crafted for Wine Production

Jana RUDOLF, Marguerite DOLS-LAFARGUE, Claire LE HENAFF-LE MARREC and Patrick LUCAS
OENO, INRAE, Villenave-d'Ornon, France

8.1. Introduction

Among lactic acid bacteria, the *Oenococcus oeni* species is the best adapted to wine. It carries out malolactic fermentation (MLF), a secondary transformation that improves the taste and aromatic complexity of wine (Peynaud 1980). *Oenococcus oeni* and MLF have only been known and mastered since the second half of the 20th century, but it is likely that they have accompanied wine production since the beginnings of winegrowing, some 8,000 years ago (Figure 8.1). Is *O. oeni* simply adapted to wine, or has it been domesticated? The domestication of microorganisms is more difficult to understand than that of animals or plants. Unlike the latter, the very existence of microorganisms has long been ignored, and they have generally been domesticated in an involuntary manner (Steensels et al. 2019). For example, during the production of bread dough, a portion of the day's fermented dough is kept to start a new batch the following day. By repeating this process over a long period, the microbial strains that are present in the dough, separated from their "wild" congeners, evolve by becoming better adapted to the fermenting environment. They become dependent on it and are domesticated.

Genetics of Domestications,
coordinated by Georges PELLETIER.

Genetic modifications in domesticated microorganisms have been detectable for about 10 years now, thanks to the development of new DNA sequencing and genome analysis technologies. Nevertheless, the hypothesis of *O. oeni*'s adaptation to wine runs up against several contradictions. Firstly, while most food production takes place throughout the year, wine MLF only lasts a few weeks a year. As a result, *O. oeni* only proliferates for a short period of time in wine, reducing its potential for evolution and domestication. Second, *O. oeni* and MLF have only been desired and controlled since the 1950s–1960s, which raises the question of the role of human activity in the domestication of the bacterium. Lastly, until the 1950s, it was strains of wild *O. oeni* that carried out MLF in all wines, and today they are still used to produce the majority of wines. These wild strains are naturally present in vineyards, on the surface of grapes and in wineries, but they can change from year to year, which is not favorable to the isolation and evolution of specific lineages. All of these particularities are, a priori, hardly compatible with the hypothesis of *O. oeni* domestication. However, genomic analyses carried out over the last 10 years confirm that this bacterium has been domesticated. This chapter first presents the particularities of the *O. oeni* species, then describes the genetic characteristics associated with its domestication.

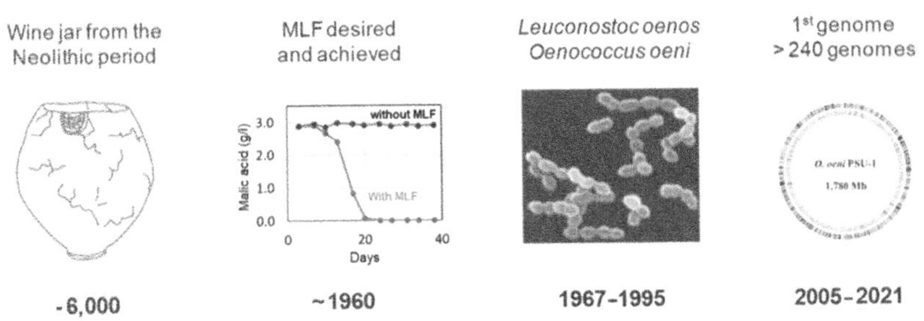

Figure 8.1. *Key stages in the description of O. oeni and MLF (source: P. Lucas). For a color version of this figure, see www.iste.co.uk/pelletier/genetics.zip*

COMMENT ON FIGURE 8.1.– *The bacterium was probably present in wine from the very beginnings of winegrowing, with the first traces dating back to 6,000 BCE. It develops spontaneously in wine and carries out MLF, which has only been known and mastered since the 1950s–1960s (Peynaud 2012). The species was first described in 1967, but was eventually renamed in 1995. The first O. oeni genome was sequenced in 2005. Today, over 300 strains have been sequenced.*

8.2. *Oenococcus oeni*, a wine bacterium for MLF

8.2.1. *A lactic acid bacterium*

Oenococcus oeni belongs to the group of lactic acid bacteria, which includes several hundred other species. Lactic acid bacteria occupy numerous ecological niches associated with plants, animals and humans. They are also used to produce thousands of traditional fermented foods, such as dairy products, cheeses, cold meats, bread, sauerkraut, wine, cider and Asian sauces (Drider and Prévost 2009). Some species thrive in a variety of ecological niches, while others, such as *O. oeni*, are more specialized. *O. oeni* exclusively proliferates in wines, ciders and kombucha, which are three products characterized by high acidity (pH 2.5–4.5), the presence of ethanol (~1–16%) and polyphenols. The main characteristic of lactic acid bacteria is that they ferment glucose and other sugars, mainly producing lactic acid and, occasionally, acetic acid and carbon dioxide. This ability of *O. oeni*'s to metabolize a high amount of sugars in wine is undesirable as this can generate excess acetic acid, which gives an unpleasant sour taste, characteristic of "vinegar-tainted" wines. Lactic acid bacteria also metabolize organic acids, amino acids and many other compounds. Thanks to these metabolic activities, they modulate taste, aroma and texture of the fermented foods in which they thrive.

8.2.2. *The wine bacterium*

Oenococcus oeni is "the" wine bacterium. It is not the only bacterium that can be detected in wine, but it is by far the most important, both in terms of abundance and the role it plays during winemaking. It is, however, a minor species in the natural environment. It is difficult to detect on grapes, whereas several hundreds or thousands of other lactic acid bacteria can be present on the surface of a single grape. When the grapes are harvested and put into vats, the bacteria begin a transition from a nutrient-poor, neutral pH, air-exposed, adherent state to a liquid, sugar- and nutrient-rich, oxygen-free, highly acidic environment (pH ± 3.5), which may contain inhibitors such as phenolic compounds or sulfites. Grape must is a hostile environment, in which many microorganisms cannot survive. Nevertheless, a few species of lactic acid bacteria thrive there, including *O. oeni*. Then, during alcoholic fermentation, the yeast *Saccharomyces cerevisiae* transforms almost all fermentable sugars into ethanol, which is a powerful antimicrobial. Acidity and increased ethanol content lead to the disappearance of almost all lactic acid bacteria, with the exception of *O. oeni*, which is highly resistant to these conditions. A few Lactobacillus and Pediococcus species can also survive and gradually grow in wine if it is not too acidic (pH > 3.5). *Oenococcus oeni* therefore takes advantage of this competition-free environment to multiply during and/or after alcoholic fermentation.

At the end of fermentation, it is almost always the only species of lactic acid bacteria that can be detected in wine (Ribéreau-Gayon et al. 2017).

8.2.3. *The MLF bacterium*

MLF starts when the *O. oeni* population reaches a sufficiently high level for the impact of the bacterial metabolism to be perceptible, in other words, at least 1 million cells per milliliter. This mainly involves the transformation of the wine's malic acid into lactic acid and carbon dioxide. The disappearance of malic acid reduces the acidity of the wine, giving it a sweeter, more balanced taste. During MLF, the aromatic complexity of the wine is also modified. The increase or decrease in "fruity", "milky", "woody" or "vegetal" notes is linked to the various metabolic activities of *O. oeni*, which transform numerous wine compounds. The value of carrying out MLF to improve the taste of wine was only understood in the 1920s–1930s, and it is only since the 1950s–1960s that it has been systematically carried out during the production of red wines, as well as some white wines (Peynaud 2012). Winemakers have adapted their winemaking practices to allow lactic acid bacteria to develop following alcoholic fermentation, and thus achieve MLF. Their efforts generally consist of maintaining the wine at a temperature close to 20°C, so as to encourage bacterial growth, as well as delaying the addition of sulfur dioxide to the wine until MLF is complete. Sulfur dioxide preserves wine by eliminating microorganisms and prevents MLF from taking place if used before completion. Since the 1990s, MLF have also been started by inoculating selected strains of *O. oeni*. Today, over 50 strains are commercially available. Nevertheless, the vast majority of MLFs are still carried out spontaneously, relying exclusively on the development of wild strains. Lastly, although MLF has only recently been mastered, many wines have probably always undergone this process, albeit randomly and more or less completely. Indeed, the analysis of wines produced before 1950 shows that they sometimes had low or zero malic acid levels, which is characteristic of partial or complete MLF (Ribéreau-Gayon et al. 2017). Thus, *O. oeni* has most certainly accompanied wine production from its very beginnings.

8.2.4. *Description of the species O. oeni*

Bacteria of the *O. oeni* species were likely isolated and analyzed for the first time at the beginning of the 20th century. They were named *Micrococcus malolacticus*, in reference to their small size, spherical or ovoid shape and ability to perform MLF. The species was then described in 1967 and named *Leuconostoc oenos*, due to its morphological and phenotypic similarities with other species of the genus *Leuconostoc*, and its predominance in wines during MLF. In 1995, the first

phylogenetic analyses showed that these *L. oenos* belong to a genetic lineage that is far removed from the other bacteria of the genus *Leuconostoc*. The species was therefore renamed *O. oeni* and classified in a new genus, *Oenococcus*, in which it has remained as the only known species for a long time. Since then, three new species of the genus *Oenococcus* have been described: *Oenococcus kitaharae*, first isolated in 2006 from distillation residues of shochu, a Japanese spirit; *Oenococcus alcoholitolerans*, isolated in 2014 from Brazilian bioethanol and cachaça production; and *Oenococcus sicerae*, isolated in Europe in 2019 and 2020 from ciders and kefir. Remarkably, all four *Oenococcus* species evolve in different products, which are acidic and contain varying degrees of ethanol.

Since the 1960s, numerous teams have contributed to the isolation and analysis of *O. oeni* strains. This work has confirmed that the species is predominant in all wines, but that it is also abundant in ciders, where it carries out MLF, and in kombucha, where its technological interest is still uncertain. Numerous metabolic pathways and enzymes involved in the transformation of malic acid and other wine compounds have gradually been identified, along with enzymes and stress resistance mechanisms that give it a good ability to survive in wine (Ribéreau-Gayon et al. 2017). Recently, the development of DNA sequencing technologies has made it possible to analyze the population structure of the species and the genetic diversity of strains. The first genome sequence of an *O. oeni* strain was produced in 2005. Its size, around 1.78 million base pairs (corresponding to around 1,700 protein-coding genes), is quite small when compared to other lactic acid bacteria species (up to around 2,800 genes). Another feature is that it contains just two copies of the operon[1] encoding the 5S-16S-23S ribosomal RNAs[2], whereas four to nine copies are generally present in other lactic acid bacteria. In general, the higher this copy number, the faster the growth rate of bacteria, as the synthesis of a greater number of ribosomes would enable the bacteria to adjust their level of protein synthesis more rapidly in a fluctuating environment. In line with this hypothesis, *O. oeni* is known to grow slowly, and is a minor species in the natural environment, where it appears to be uncompetitive with other lactic acid bacteria. Over the past 10 years, high-throughput sequencing methods have made it possible to compare the genomes of more than 240 strains. They range in size from 1.7 to 2.2 million base pairs[3], or

1 A group of contiguous genes that operate under the signal of the same *promoter*, a region of DNA that controls their ***transcription***.

2 Ribonucleic acids are the building blocks of ribosomes, the cellular organelles that are responsible for ***translating*** messenger RNA into proteins.

3 The length of DNA molecules is counted by the number of matched base pairs located on the two complementary strands of the double helix.

around 1,700 to 2,200 genes, demonstrating significant intraspecific genetic diversity. The "core genome" of the species, meaning all of the genes or gene families common to all strains, comprises 900 genes, in other words, around half the genetic content of each strain. The remaining genes are present in only one or several strains, but not in all. Thus, to date, over 9,500 different genes or gene families making up the ***pan-genome*** have been identified in the 240 available genomes, highlighting the species' significant genetic diversity (Campbell-Sills et al. 2017, pp. 457–468).

8.3. Genetic characteristics of *O. oeni* domestication

8.3.1. *Characteristic traits of domesticated microorganisms*

According to Diamond (2002), a domesticated species is "a species bred in captivity and thus modified away from its wild ancestors, so as to make it more useful to humans who control its reproduction and (in the case of animals) its diet". This definition obviously applies to farm animals, domestic pets and cultivated plants, but it also concerns microorganisms in fermented foods and beverages. Food production creates new environments in which microorganisms can evolve, adapt and specialize. They acquire the capacity to resist specific stress factors related to these environments, as well as the ability to consume the available nutrients. At the same time, this specialization leads to a loss in capacities that were useful in the original natural environment, but which are sometimes redundant in the food environment. Thus, the "fermenting" strains evolve to form new genetic lineages that are distinct from "wild" strains. Ultimately, humans can also influence the evolution of domesticated strains by selecting the fermented products that best suit their tastes. In this way, man selects strains with the most appreciated metabolic capacities. The domestication of microorganisms is therefore most often "passive", or involuntary. It differs from the "active" domestication of plants, and especially animals, where man selects individuals from generation to generation. Nevertheless, the domestication of microorganisms is linked to human activity. The phenotypic and genomic changes that occur during this domestication are beginning to be identified in yeasts, bacteria and molds used in various food productions. Despite the phylogenetic distance between them, domesticated microorganisms share many characteristics (Gallone et al. 2020):

– They are genetically distinct from wild lines, which limits gene exchanges with the latter.

– They are dependent on or partially specific to the environment created by man.

– By adapting to a single type of environment, their genetic diversity is more limited than that of wild lineages.

– They show signs of genomic reduction, in other words, loss of nonessential genes in the artificial environment in which they have been domesticated.

8.3.2. *Identification of domesticated O. oeni genetic lines*

Since the 1960s, thousands of *O. oeni* strains have been isolated from numerous wine-producing regions. Various typing methods have been used to differentiate them, and some are still employed today, such as rapid amplification of polymorphic DNA by *PCR* or MLVA (analysis of multiple ***loci*** containing variable genetic regions). However, it was not until the use of DNA sequencing-based techniques, such as multiple locus sequencing typing (***MLST***), that we began to see the phylogenetic links between strains. In 2009, MLST analysis of some 50 *O. oeni* strains from different regions, and different wines and ciders, revealed the existence of at least two genetic lineages, grouping strains isolated from wine (lineage A), and those that are sometimes isolated from wine, but more frequently from grape must and cider (lineage B). Since then, the development of high-throughput DNA sequencing technologies has made it possible to sequence more and more entire strain genomes, and to also use them to measure the phylogenetic distribution of strains with great precision. Today, more than 240 genomes from strains of different origins are known. Their analysis suggests that there are at least four major strain lineages (Figure 8.2):

– lineage A: strains isolated from wines;

– lineage B and lineage C: strains mainly isolated from grape must and cider;

– lineage D: strains isolated from kombucha.

The detection of *O. oeni* in kombucha is very recent. The exclusive presence of lineage D strains in this product is not yet explained, but it is likely that the specific physicochemical parameters of kombucha favor the growth of strains of this lineage and inhibit others. Indeed, kombucha contains very little ethanol (±1%) compared with cider (±5%) and wine (±12%). However, it can be significantly more acidic, with a pH of around 3.3–3.5, and sometimes below 3.0 or even 2.0, compared with ±3.5 for wine and ±4.0 for cider. Kombucha is also very rich in polyphenols, which are different from those in wine or cider.

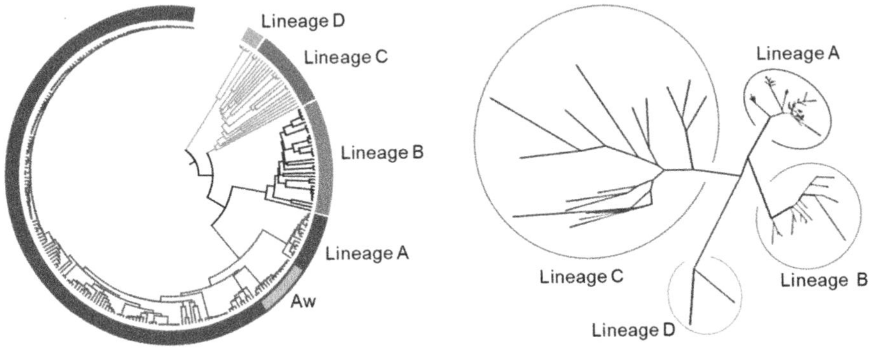

a) b)

Figure 8.2. *(a) Phylogenetic distribution of 226 O. oeni strains in four major lineages (A–D) and various sub-lineages such as "Aw", a sub-lineage that is specific to low pH white wines; (b) phylogenetic tree reconstructed using the maximum likelihood estimation, showing the evolutionary distances between the four A–D lineages (source: P. Lucas). For a color version of this figure, see www.iste.co.uk/pelletier/genetics.zip*

Lineages A, B and C are detected in grape must and cider. In grape must, populations of strains B and C are significantly higher than those of lineage A, but this situation is reversed during alcoholic fermentation. With increasing ethanol content, the occurrence of lineages B and C declines more or less completely, while lineage A strains survive, thrive and are virtually the only ones detectable during MLF. According to phylogenetic analyses, all strains of lineage A are much more closely related to each other than to strains of other lineages (Figure 8.2(b)). This genetic proximity is characteristic of a microbial lineage that has evolved across an ecological barrier. All lineage A strains probably derive from an ancestral strain characterized by a better ability to grow in wine. In addition, several A "sub-lineages" can be detected (Figure 8.2(a)). One of these includes strains that are exclusively found in white wines, which usually have the lowest pH levels (pH ±3.0, Aw lineage). These are white wines from Burgundy, Champagne and the Loire Valley, suggesting that the strains are not specific to their regions of origin, but are adapted to the types of products in which they develop. Indeed, these Aw strains are more resistant to the acidity of these white wines than other *O. oeni* A strains, but they are clearly more sensitive to the presence of polyphenols, which are more abundant in red wines. Phylogenetic analyses therefore suggest that lineage A has crossed an ecological barrier by becoming able to develop in wine after alcoholic fermentation, and that sub-lineages, such as Aw, have crossed a second ecological barrier by acquiring greater adaptability to certain types of wine. This specialization no longer enables them to develop efficiently in grape must or other wines, which is typical of domesticated organisms.

8.3.3. *Genetic mechanisms that contribute to the domestication of O. oeni*

8.3.3.1. *Hypermutability (absence of mutS-mutL genes)*

O. oeni is a "hypermutating" or "fast-evolving" species. This characteristic was detected as early as 1989, when the first phylogenetic tree that positioned *O. oeni* among other lactic acid bacteria was constructed. Its position at the end of a very long branch suggested that it had evolved much more rapidly than other species. The sequence of the first *O. oeni* genome (PSU-1 strain) enabled us to identify the genetic characteristic that would be responsible. In fact, *O. oeni* is one of the few species lacking the universal DNA damage repair system[4] encoded by the *mutS* and *mutL* genes. Their absence greatly increases the frequency of point mutations, recombination events, exogenous DNA integration by ***horizontal transfer*** and, ultimately, the evolution of the bacterial genome. In *O. oeni*, the spontaneous mutation rate is 100–1,000 times higher than in other lactic acid bacteria. It has been suggested that the loss of *mutS-mutL* genes is a consequence of *O. oeni* domestication. However, they are absent in all *O. oeni* strains, while only lineage A shows wine domestication characteristics. They are also absent in the three other *Oenococcus* species that are not found in wine: *O. kitaharae*, *O. alcoholitolerans* and *O. sicerae*. The absence of *mutS-mutL* genes is therefore characteristic of bacteria of the genus *Oenococcus*. This is not a consequence of their domestication, but it has undoubtedly favored their genetic evolution and domestication in the various products in which they proliferate.

8.3.3.2. *Horizontal gene transfer*

Horizontal gene transfer, or "non-genealogical transmission of genetic material from one organism to another", plays a key role in the evolution of bacteria. The integration of exogenous DNA enables the acquisition of new genes and functions that may be useful for adapting to, or changing, the ecological niche. It also favors genomic reshuffling and gene loss, which can be beneficial, neutral or harmful to the host cell. Horizontal gene transfer is common among lactic acid bacteria in food environments where multiple species and strains coexist. It occurs through three independent mechanisms:

4 Presence of noncomplementary bases between the two DNA strands. The complementary bases that ensure the pairing of the two strands of the double helix are adenine with thymine and guanine with cytosine.

– Conjugation, which enables DNA transfer when there is direct contact between a donor cell and a recipient cell. It is generally mediated by so-called "conjugative" plasmids[5] or integrative conjugative elements (ICEs), which combine the characteristics of ***transposable elements***, conjugative plasmids and bacteriophages[6]. They are conjugative, because they encode all of the proteins needed to form a protein channel between the two cells, regulate DNA transfer and potentially integrate it into the host cell chromosome.

– Transduction (generalized, specialized or lateral), by which new genes, chromosomal rearrangements and changes in gene expression occur in bacteria during their interaction with bacteriophages. Indeed, temperate phages, in other words, those that can integrate into the DNA of the cell they infect, play a key role in the evolution of bacterial genomes and the regulation of gene expression.

– Natural transformation, which occurs in naturally competitive bacteria, in other words, those with enzymatic systems capable of capturing exogenous DNA – for example, from a dead cell – and transferring it into the cell and integrating it into the chromosome.

Various genetic signatures can be used to identify genes acquired by horizontal transfer. These include, for example, the proximity of transposase genes[7], insertion sequences (IS) (see below) or phage genes, sequences that show strong identity with genes from other species, or biases in base composition relative to the rest of the genome. In *O. oeni*, there is ample evidence of genes or genomic regions acquired by horizontal transfer.

8.3.3.3. *Plasmids, insertion sequences and regions of plasticity in O. oeni*

Most of the plasmids described to date in *O. oeni* are small (3,000–5,000 base pairs or kb) and only encode a few of the proteins required for replication. Only one medium-sized plasmid has been sequenced (pOENI-1, 20 kb). It is more frequent in strains that are well adapted to wine. This prevalence is probably due to one of its genes encoding a "sulfite pump" (see section 8.3.4.5). Larger plasmids have been detected in various strains, but none have been isolated and sequenced. This may be due to the instability of larger plasmids when bacteria are cultured in the laboratory.

5 Autonomous DNA molecule, generally circular, distinct from the bacterial genome and not essential to the cell's survival.

6 Or phage, a virus that infects bacteria.

7 Enzyme encoded by a transposable element that allows it to be moved from one locus to another in the genome.

To date, no functional ICE-type conjugative ***transposons*** have been described in *O. oeni*. However, other mobile genetic elements, known as IS, are very frequently found. IS are small genetic elements (~600–1,600 base pairs) encoding a transposase and flanked by short DNA repeats. In *O. oeni*, they are divided into seven different families (IS30, IS3, IS5, ISL3, IS6, IS256 and IS982) shared with other species of lactic acid bacteria. IS contribute to genome evolution by promoting the mobility of adjacent genes. In *O. oeni*, they may have favored the acquisition of various genes linked to adaptive value and domestication. These entities can also be inserted into other variable regions of the genome. The mechanisms are complex and often involve successive stages of integration at a single site, like a Russian doll. These accretion mechanisms create large regions of plasticity within genomes.

8.3.3.4. *O. oeni phages*

Phages play a major role in the evolution of bacterial genomes. Indeed, temperate phages can insert a copy of their genome (prophage) into the bacterial chromosome, thereby contributing genes and functions of interest to the lysogenic bacterium (which contains a prophage that, under the effect of an external factor, can generate virulent bacteriophages and cell lysis), or modulate the expression of its genes. The presence of prophages in the genomes of *O. oeni* strains has been known since the 1990s. A recent inventory of 231 strains revealed 187 complete prophages in 134 strains (58%). Most of the 134 lysogenic strains contain a single prophage, but 43 have two or three. Prophages range in size from 35 to 46.2 kb, representing 2–6.7% of bacterial DNA. They belong to four genetic groups (A–D), defined according to the sequences of their integrases (enzymes that enable their integration into the bacterial genome), as well as attachment and insertion sites. Incomplete or degenerate prophages, ranging in size from 5.2 to 19.5 kb, are also present in almost all strains. They form five groups that are defined according to the five insertion sites in which they are found. These prophage remnants have lost most of their important structural or functional genes, but their persistence in the genomes means that they may still be useful to the bacterium. For example, they encode surface proteins of unknown function, as well as restriction-modification systems that probably give the bacterium increased resistance to superinfection by other phages. The distribution of complete and incomplete prophages varies between *O. oeni* lineages and genetic sub-lineages. All strains in lineage A contain the degenerate "PR_C" prophage, while only nine out of the 21 lineage C strains do, and none of the 31 B- or the five lineage D strains. Furthermore, lineage A contains the most lysogenic (63%) and polylysogenic strains (45 of 47 strains possessing two or three prophages), and the 10 sequenced strains of the Aw sub-lineage are all lysogenic, whereas less than 40% of strains from other sub-lineages are. There is no doubt that phages have contributed to the domestication of *O. oeni* in wine.

8.3.3.5. *Natural competence in O. oeni*

Natural competence enables a bacterium to integrate an exogenous DNA fragment. This is made possible by the activity of DNA membrane receptors, a transmembrane pore and proteins that transport DNA into the cell. Pan-genome analysis of 191 *O. oeni* strains has revealed the genes encoding all of these proteins. They are present in strains of genetic lineages A, B and C, but are only complete in lineages B and C. At least three of these genes encode proteins that are truncated and probably nonfunctional in all lineage A strains. This distribution could be explained by the fact that natural competence is still useful for strains of lineages B and C, in order to evolve genetically and adapt to diversified or changing environments, whereas strains of lineage A are already adapted to a simplified environment of stable composition.

8.3.4. *Genetic signatures of domestication*

8.3.4.1. *Genome reduction*

Gene loss and genome size reduction in bacteria is a common process that occurs during adaptation to a specific environment and a fortiori during domestication. Indeed, point mutations and recombination events can not only increase genetic complexity at the genome level but also reduce it by inactivating genes (formation of pseudogenes) or inducing their loss. This phenomenon of genome reduction is common to lactic acid bacteria. The common ancestor of all *Lactobacillales* is thought to have lost 600–1,200 genes (in other words, 25–30% of its total gene pool), and to have acquired fewer than 100 during its divergence from its *Bacilli* ancestor. Many of these changes are linked to the transition to rich ecological niches. Lactic acid bacteria have lost the ability to biosynthesize many cofactors, amino acids, fatty acids and other compounds available in their new environment. Conversely, they have often acquired genes for transporters, peptidases and other functions, enabling them to draw cofactors and nutrients from the environment. *O. oeni* has one of the smallest genomes among lactic acid bacteria. It is characterized by a large number of lost genes, which is consistent with its specialization in a relatively stable, nutrient-rich and uncompetitive environment. Gene inactivations and losses have specifically occurred in different genetic lineages.

8.3.4.2. *Genetic variations between strains of lineages A, B, C and D*

Based on phylogenetic analyses (Figure 8.2), there can be no doubt that the strains in genetic lineage A are all descended from the same common ancestor. It is then possible to imagine that this ancestor acquired the genes and functions that gave it the ability to grow in wine, that these characteristics still persist in all strains of lineage A, and that they are absent in lineages B, C or D. The genome analyses carried out to date have highlighted the particular genetic characteristics of wine-adapted strains – several of which are described from sections 8.3.4.3–8.3.4.6. However, none are shared by all strains of lineage A, nor are they absent in all other strains of lineages B, C and D. The current diversity of strains is the result of numerous genetic changes that have occurred since the divergence of lineage A. The strains of this lineage are not adapted to the wine in which their common ancestor evolved, but have progressively adapted to the different types of wine produced today. For example, strains of the Aw lineage perform very well in acidic white wines, but are unsuitable for red wines. It is likely that other strains perform better in wines made from certain grape varieties, or richer in ethanol, polyphenols or sulfites. In addition, strains of lineages B and C are sometimes detected during or after MLF, just as it is possible to detect other species of lactic acid bacteria in wine after MLF if no action is taken to remove them (Ribéreau-Gayon et al. 2017). These bacteria are not as well adapted to wine as those of lineage A: they grow more slowly and less frequently, however, they do contain genes and functions that allow them to survive. Thus, the genetic characteristics that improve the suitability of the wine certainly vary between strains of lineage A, and some of these characteristics are probably found in strains of lineages B, C and other species of lactic acid bacteria.

8.3.4.3. *Key MLF genes: the "malolactic" operon*

During MLF, the main transformation carried out by the bacteria is the conversion of malic acid into lactic acid and carbon dioxide. This is achieved by a malolactic enzyme, a malic acid membrane transporter and a transcription regulator. The combined activities of the malolactic enzyme and the transporter facilitate the elimination of one intracellular proton (H+) for each malic acid molecule transformed. The loss of this proton enables the bacterium to regulate its intracellular pH and better survive in an acidic environment. It also enables the production of energy in the form of ATP, via ATP synthetase. The malolactic enzyme and transporter genes are found in various species of lactic acid bacteria, but generally only in a few strains of each species. In *O. oeni*, all strains from lineages A to D possess them. They are also present in *O. kitaharae*, *O. alcoholitolerans* and *O. sicerae*. Consequently, it was not the acquisition of these genes that enabled the domestication of lineage A to wine. Rather, they are a characteristic of bacteria evolving in a medium containing malic acid, such as grape, apple or other fruit must. They are probably no longer useful in *O. kitaharae*, as the malolactic enzyme gene in this species is mutated and nonfunctional. Similarly, *O. oeni* strains of lineage D, which do not grow in wine but in kombucha, probably no longer have use for the malolactic operon genes, as their transcription regulator gene is mutated and probably inactive.

8.3.4.4. *Acquiring and losing the ability to metabolize sugars*

Like all lactic acid bacteria, *O. oeni* is strictly fermentative and develops in the presence of sugars, from which it draws its energy. However, not all *O. oeni* strains use the same sugars. Lineage A strains are generally less versatile than lineage B and lineage C strains. Taking all lineages together, 75% of the strains studied metabolize glucose, trehalose, cellobiose, ribose, mannose and melibiose. Fructose and arabinose are added to these substrates, followed by maltose and sucrose, which are consumed by most lineage B and C strains, but by very few lineage A strains. Galactose, raffinose and xylose are consumed by 5%–10% of strains. The reduced catabolic capacity of lineage A strains could be a consequence of their domestication. Indeed, in wines undergoing MLF, the most consumed substrates are glucose, trehalose and sometimes cellobiose, far ahead of all other available sugars, especially pentoses. The catabolism of pentoses, maltose or sucrose would, however, be useful in other habitats that are frequently occupied by B and C strains (apple juice or other fruits).

Most sugar metabolism genes are present in all strains of the species. Loss of metabolic capacity is generally due to single mutations, leading to gene inactivation. For example, genes for fructose transport and arabinose or maltose catabolism have become pseudogenes (genes inactivated by mutations) in almost all lineage A strains. Nevertheless, gene acquisition by horizontal transfer can, in some cases, restore the capacity to metabolize arabinose (*xylA*, *xylD* and *araD1* genes) or fructose (pts^{fru} gene). No complementation restores maltose catabolism, which is very little or not at all present in wine. The loss of this metabolic capacity is therefore a consequence of the domestication of lineage A strains. Xylose catabolism, on the other hand, is only possible following the acquisition of specific genes by lineage A and B strains. This metabolism would therefore not be essential in the context of wine or fruit juices.

8.3.4.5. *Acquiring a sulfite pump*

Lactic acid bacteria often possess plasmids that provide accessory functions that can enhance their adaptive value in specific ecological niches. In *O. oeni*, only one plasmid containing accessory genes has been described to date: pOENI-1. It contains around 20 genes, including the *tauE* gene, which encodes a membrane transporter with strong sequence similarities to a sulfite transporter. In some strains, *tauE* is located on the bacterial chromosome, probably as a result of a partial integration of the pOENI-1 plasmid. The "sulfite pump" activity encoded by *tauE* could enhance bacterial survival in wine by giving it resistance to sulfites that are added to the must or produced by yeast during alcoholic fermentation. Indeed, *tauE* is present in most strains that develop after alcoholic fermentation and carry out MLF. According to the analysis of *O. oeni* genomes available to date, *tauE* is absent from lineages C and D, very rarely present in lineage B, but found in almost half of lineage A strains. Although its functional role remains to be confirmed, *tauE* has all of the characteristics of a gene that has contributed to the domestication of *O. oeni* to wine.

8.3.4.6. *Biosynthesis of exopolysaccharides (EPS) and the special case of the gtf gene in the Aw lineage*

O. oeni strains produce various types and quantities of exopolysaccharides (EPS), which can modulate their resistance to wine, improve their survival in the winery from one vintage to the next, or promote their dissemination. They possess numerous genes dedicated to the biosynthesis of homo- or heteropolysaccharides, which are localized in several chromosomal loci (Figure 8.3). This diversity is correlated with strain lineages, each possessing a different panel of *eps* genes (Table 8.1).

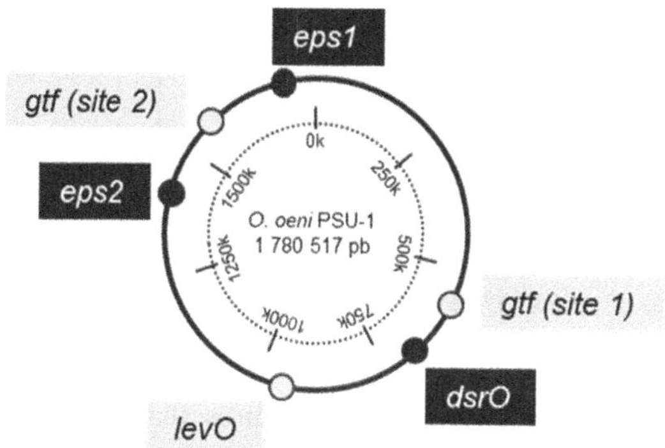

Figure 8.3. *Chromosomal localization of genes and gene clusters associated with EPS synthesis in O. oeni. The genome of the PSU-1 strain is used as a reference. Genes highlighted in dark blue are conserved in most strains. Those in light blue are only present in a few strains (source: P. Lucas). For a color version of this figure, see www.iste.co.uk/pelletier/genetics.zip*

	Sub-lineage	*eps1*	*eps2*	*levO*	*gtf*
Lineage A	S14	Type A	Type 277	Absent	Absent
	PSU-1	Type A	Type PSU-1	Absent	Absent
	B429	Type B	Type B429	Absent	Absent
	B553	Type B	Type B553	Absent	Absent
	Aw	Type C	Incomplete	Absent	Present on site 1
	VF	Type C	Absent	Absent	Absent
	277	Type B	Type 277	Absent	Absent
	0607	Type B	Type 0607	Absent	Absent
Lineage B	0502	Type A	Type 0502	Present	Present on site 2
	S13	Type A	Type S13	Incomplete	Absent
	9805	Type A	Type 9805	Present	Absent
	BAA1163	Type A	Type 1163	Present	Absent
	9304	Type A	Type 9304	Present	Absent
	0501	Type A	Type 0501	Present	Absent
Lineage C	C52	Type A	Type C52	Present	Absent

Table 8.1. *Distribution of major exopolysaccharide production genes and gene clusters in lineages and sub-lineages of O. oeni strains*

Lineage A strains have a number of features that could contribute to their domestication.

First, their *eps2* locus is different. In fact, this locus is one of the most variable regions of the chromosome, differing in every strain lineage. The very high diversity of *eps2* genes is reminiscent of the *cps* (capsular polysaccharide) loci of *Streptococcus pneumoniae*, in which the polysaccharide capsule is a key factor in virulence. However, the *eps2* locus is also variable within lineage A, with each sub-lineage A possessing at least one different gene panel. Thus, a role of *eps2* in *O. oeni* domestication is possible, but remains to be demonstrated.

Second, strains of lineage A differ from those of lineages B and C in the absence of the *levO* gene. This gene is involved in the synthesis of levan from sucrose, which is thought to enhance the bacteria's ability to catabolize sucrose. As sucrose is absent from wine during MLF, there is no selective pressure on this gene, which could explain its disappearance from lineage A and how it could have contributed to its domestication.

Third, within lineage A, strains of the Aw sub-lineage are highly unusual. In particular, they have an incomplete *eps2* locus, consisting solely of pseudogenes, and a *gtf* gene that is not present in any other lineage A strain, and only in one lineage B strain. This *gtf* gene is inserted close to prophage genes, but not in the same genomic regions in Aw and B strains (Figure 8.3), suggesting that it was acquired by independent horizontal transfer events. This mobile gene is also known from other species of lactic acid bacteria, where it has been found in the sequences of four plasmids. The *gtf* gene encodes a transmembrane glucosyl transferase, which polymerizes glucose by forming β-1,3 and β-1,2 branches, releasing the EPS formed in the extracellular space. Some of this EPS remains in the cell microenvironment, without forming a real capsule. However, its presence gives bacteria the capacity to adhere, form biofilms and resist stress. Indeed, the bacteria that produce it have a survival rate 100 to 1,000 times higher when grown at very acidic pH (2.7) or in white wine. Since strains of the Aw sub-lineage are particularly resistant to acidity and are found almost exclusively in white wines, the acquisition of *gtf* undoubtedly represents a secondary domestication event to this type of wine. The organization of the *eps* genes in the Aw sub-lineage is reminiscent of that found in *Streptococcus pneumoniae* type 37, a singular streptococcal lineage that lacks a *cps* operon, but has acquired the *tts* gene, which encodes a protein with an activity very similar to that encoded by *gtf*. The domestication of the Aw lineage to high-acidity white wines thus occurred with a radical change in the type of EPS produced by the bacteria.

8.4. Conclusion

Our knowledge of the domestication of *O. oeni* is very recent. It has been acquired because of new DNA sequencing technologies and computer tools developed over the last 10 years. To date, just over 200 strains of *O. oeni* have been sequenced, which is far from enough to clarify all of the genetic changes that accompanied the domestication of the species to wine and certain types of wine. In the years to come, many more strains from different regions and wine types will have to be analyzed to understand the evolution of this species' lineages, as well as understand how human activity has promoted their emergence.

Nevertheless, the data currently available confirm that this species has indeed been domesticated for wine. In particular, the existence of lineage A, which represents by far the majority of *O. oeni* in wine and is genetically distinct from lineages B, C and D, which are better adapted to less alcohol-rich environments. Lineage A is in the minority in grape must and cider, where lineages B and C proliferate better, and is absent from kombucha, where only lineage D has currently been detected. There can be no doubt, therefore, that this lineage has undergone a process of passive domestication, during which it has acquired the capacity to survive and proliferate in wine. This capacity separated it from other lineages and enabled it to continue evolving, adapting to the presence of polyphenols, sulfites and other substrates still present after alcoholic fermentation. On the other hand, it has become less competitive in less alcohol-rich environments and probably in the natural environment. Indeed, lineage A strains often have a lower growth rate than those of other lineages when grown in ethanol-free media. Future studies will need to explain the capacities acquired, but also those lost during domestication.

To date, the genetic changes that occurred during the domestication of *O. oeni* to wine are still largely unknown. The acquisition of the *tauE* gene by a majority of lineage A strains probably gave improved resistance to sulfites. Remarkably, a similar phenomenon has occurred in the oenological strains of the *Saccharomyces cerevisiae* yeast, which have acquired a "sulfite pump" encoded by the *SSU1* gene. The widespread use of sulfites in wines over the last few centuries has thus promoted the evolution of all microorganisms that were already adapted to wine. The acquisition of these genes is the result of secondary domestication events, as is the acquisition of the *gtf* gene, which gives Aw strains greater resistance to acidic white wines. These genes provide an additional advantage but are not at the origin of the adaptation of lineage A to wine. Numerous genes for resistance to acidity or other types of stress have already been identified in *O. oeni*, including genes for the metabolism of malic acid and other amino or organic acids, but most of these are not specific to lineage A (Ribéreau-Gayon et al. 2017). We also need to identify the other genes and gene expression regulation mechanisms that favored the transition of this lineage from the natural environment to wine. The evolutionary mechanisms involved also need to be identified. The absence of the DNA repair system encoded by the *mutS-mutL* genes certainly played an important role, but mobile elements such as phages and plasmids also played a key role in the evolution of the species. Ultimately, beyond scientific knowledge, this information will be essential to help winemakers – who have domesticated *O. oeni* – choose and use the most suitable strains for MLF in different types of wine.

8.5. References

Campbell-Sills, H., Lorentzen, M., Lucas, P.M. (2017). Genomic evolution and adaptation to wine of *Oenococcus oeni*. In *Biology of Microorganisms on Grapes, in Must and in Wine*, König, H., Unden, G., Fröhlich, J. (eds). Springer International Publishing, Berlin.

Diamond, J. (2002). Evolution, consequences and future of plant and animal domestication. *Nature*, 418, 700–707.

Drider, D. and Prévost, H. (2009). *Bactéries lactiques : physiologie, métabolisme, génomique et applications industrielles*. Economica, Paris.

Gallone, B., Steensels, J., Verstrepen, K.J. (2020). Moulded by humans: The domestication of blue-veined cheese fungi. *Mol. Ecol.*, 29, 2517–2520.

Peynaud, É. (1980). *Le goût du vin*. Dunod, Paris.

Peynaud, É. (2012). *Le vin et les jours*. Dunod, Paris.

Ribéreau-Gayon, P., Dubourdieu, D., Donèche, B., Lonvaud, A. (2017). *Traité d'œnologie*. Dunod, Paris.

Steensels, J., Gallone, B., Voordeckers, K., Verstrepen, K.J. (2019). Domestication of industrial microbes. *Curr. Biol.*, 29, R381–R393.

9

Tracing the Origins of Wheat Cultivation

Caroline PONT and Jérôme SALSE
GDEC, INRAE, Clermont-Ferrand, France

9.1. The different types of wheat, one of the most widely consumed cereals in the world

Flowering plants, or angiosperms, diverged over 200 million years ago into two main families, the monocotyledons and the dicotyledons. The monocotyledon family comprises the sub-family *Poaceae*, which includes most modern cereals, such as rice (sub-family *Oryzoideae*), corn, millet and sorghum (sub-family *Panicoideae*), as well as oats and *Triticeae* (sub-family *Pooideae*). Wheat belongs to the *Triticeae* botanical family, which also includes barley and rye.

Wheat, the world's most widely consumed cereal, is grown in a wide range of environments, from low-humidity regions such as Nigeria, Australia, India and Egypt, to very wet areas in South America, or cold regions in Canada. Wheat production has increased considerably over the last hundred years, reaching 836 million tons (Mt) in 2020, compared with 100 Mt in the early 1900s. The main producers are the European Union, China, the "Black Sea countries" (Russia, Ukraine, Kazakhstan), India and the United States (Figure 9.1). Direct human food consumption in the form of semolina or flour and their derivatives (bread, cookies, pasta, etc.) accounts for most of wheat use, followed by animal feed and, lastly, industrial use (starch and gluten mills).

Genetics of Domestications,
coordinated by Georges PELLETIER.

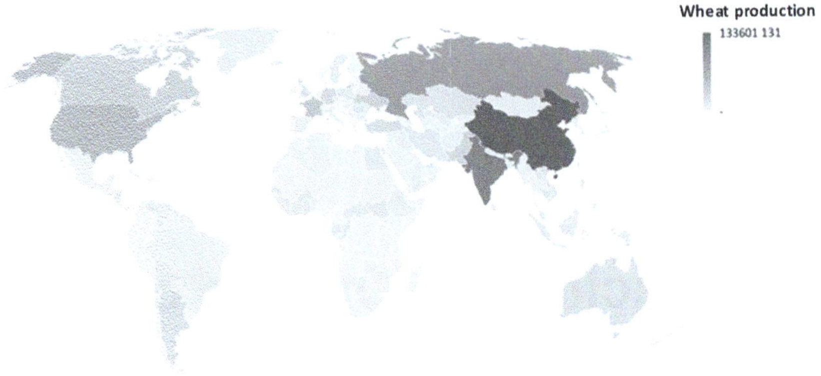

Figure 9.1. *Global distribution of wheat production in 2019, expressed in tons (scale on the right) (source: FAOSTAT). For a color version of this figure, see www.iste.co.uk/pelletier/genetics.zip*

9.2. A species with ancient origins resulting from multiple hybridizations

Soft wheat (*Triticum aestivum*), which is used to produce flour and bread, and hard wheat (*Triticum durum*), which is used to produce semolina and pasta, are the result of a complex, ancient evolutionary process that involves a unique genetic process: ***polyploidy***. A ***diploid*** species has one set of homologous chromosomes[1] (symbolized by the 2n parameter); a polyploid species has more than one complete set of homologous chromosomes; in other words, ***tetraploid*** for two complete sets of homologous chromosomes (4n) and hexaploid for three complete sets of homologous chromosomes (6n). There are two mechanisms at the origin of this evolutionary process: autopolyploidy, which results from the multiplication of the genome of a single individual, and allopolyploidy, which results from the addition and maintenance of two parental genomes following crosses between two individuals, each supernumerary genome being entirely conserved in both cases.

The schematic model proposed in Figure 9.2 (according to Pont et al. (2019a)) illustrates the evolutionary pathway that is widely accepted in the scientific community, leading to modern polyploid wheats from hybridization of diploid species by allopolyploidy.

1 Chromosomes that pair up during meiosis and exchange genetic material through recombination. They have the same structure but differ in the alleles they carry.

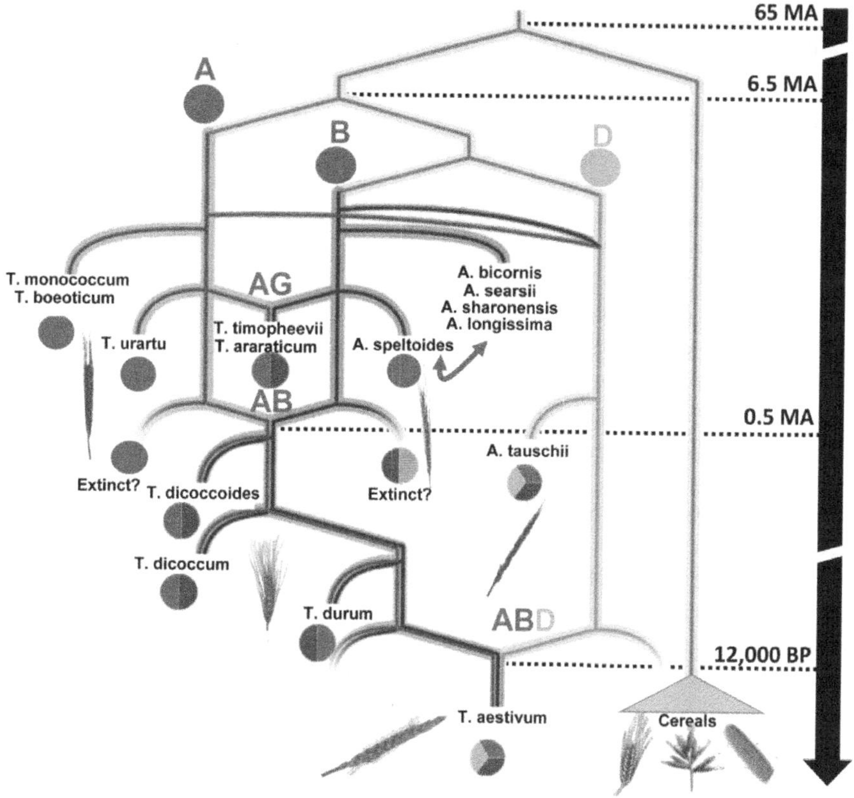

Figure 9.2. *Evolutionary model of modern wheat. For a color version of this figure, see www.iste.co.uk/pelletier/genetics.zip*

COMMENT ON FIGURE 9.2.– *Illustration of the paleo-history of wheat from diploid ancestors AA (green), BB (purple) and DD (brown) along a time scale expressed in millions of years ago (MA) and years before present (BP) (right). Sub-genomes are shown in different colors, so that hybridization events can be highlighted through a mixture of colors on the evolutionary tree. Wheat species names are shown on the evolutionary tree as referenced in the text (source: adapted from Pont et al. (2019a)).*

The diploid genome of each of these founder species, comprising seven pairs of chromosomes, is designated by a double capital letter. Therefore, wild diploid ancestors who gave the AA and SS genomes (close to BB) hybridized with each other, leading to the wild tetraploid species AABB (500,000 years ago), which then hybridized with another wild diploid species DD (almost 10,000 years ago), which gave the hexaploid *T. aestivum* lineage (AABBDD). This model suggests that the BB genome of modern wheat is derived from a lineage of the *Sitopsis* section, which gave rise to the current *Aegilops speltoides* (SS), while the progenitors of *Aegilops tauschii* and *Triticum urartu* represent the admitted origins of the DD and AA genomes, respectively. It is also accepted that the DD ancestor, which gave rise to hexaploid wheats (AABBDD) from tetraploids (AABB), is itself the result of hybridizations between AA and BB progenitors, then hybridizing again with a specific DD progenitor (El Baidouri et al. 2017). Einkorn wheat (*Triticum monococcum*) is the only known domesticated form of diploid (AA) progenitors. *Triticum araraticum* (also known as *T. araraticum* Jakubz) represents the closest wild descendant of the tetraploid AAGG ancestor. This appears to have been subsequently domesticated to form *Triticum timopheevii* (Zhuk), while hybridizing with *Triticum boeoticum* to lead to the hexaploid *Triticum zhukovskyi* AAAAGG line (Menabde and Ericzjan).

The model confirms that the wild emmer (*Triticum dicoccoides*) is the closest descendant of the ancestor of the modern AA and BB sub-genomes of the modern AABB tetraploid and AABBDD hexaploid descendants. During the first phase of wheat domestication, a pool of wild emmer plants *T. dicoccoides* (*Körn. ex Asch.* and *Graebner*) *Schweinf.* gave rise to at least two distinct lineages of domesticated tetraploids, *Triticum dicoccum Schrank ex Schübl.* (domesticated emmer, also known as *T. dicoccon* Schrank) and *T. durum Desf.* (domesticated durum wheat). Ultimately, the model supports the hypothesis that *T. aestivum* (soft hexaploid wheat AABBDD) is most likely derived from an ancestral hybridization event between *T. durum* and a DD lineage close to wild *A. tauschii*. Subsequently, *Triticum spelta* emerged from hybridization between *T. aestivum* and *T. dicoccum* (tetraploid).

9.3. Archaeological evidence of its origins: archaeobotany

9.3.1. *First traces*

When was wheat first cultivated? The first traces of farming were found in Israel, at the Ohalo II site in northern Galilee, some 23,000 years ago (Weiss et al. 2008). Traces of plant cultivation were found at this site, where a community of hunter-gatherers lived. Wild emmer, wild barley and wild oats can be found mixed with weeds, which are real markers of the existence of agricultural fields, having been unintentionally mixed with wheat during harvesting (Piperno et al. 2004).

The first evidence of a bread-like food preparation based on cereal seeds comes from the late Paleolithic Jordanian site of Shubayqa 1, within which the remains of an ancient wafer of over 14,400 years old were found (Arranz-Otaegui et al. 2018).

Lastly, at the Jerf al-Ahmar site in the Middle Euphrates region of Syria, numerous remains (grindstones, cereal patties and charred plants) attest to the importance of barley and wild wheat from 9,500 to 8,700 BCE (Willcox and Stordeur 2012).

At this stage, we cannot really speak of domestication, but of farmers cultivating wild plants. Wild cereal farming, including wild einkorn (*T. boeoticum*) and wild emmer (*T. dicoccoides*) for wheat, has been documented throughout the Fertile Crescent, central and eastern Anatolia, the Zagros mountains on the border between Iraq and Iran, as well as the Middle Euphrates region and the Levant. A little later, at the beginning of the *Holocene era*, we find the first signs of domesticated plants in Early Neolithic farming villages (Hopf et al. 2018) dating from 10,500 to 10,100 years before present (BP); this is referred to as the "neolithization" of agriculture.

Indeed, archaeobotanical remains of domestication have been found in Southwest Asia, with Emmer and Einkorn spikelet bases bearing disarticulation scars, a characteristic of a threshing harvest. In fact, this scarring attests to the fact that these cereals were already domesticated at that time. Wild wheat has a fragile rachis, which leads to spontaneous disarticulation of the spike (resulting in grains being scattered on the ground). Domesticated wheat, with its solid rachis, is harvested through threshing, enabling the grains to be collected and reducing loss of production. The latter produces a characteristic scar at the base of the spikelet (Figure 9.3).

9.3.2. *Domestication center(s)*

9.3.2.1. *A slow domestication process*

The Abu Hureyra site in Syria, where a majority of wild grains were found from 12,200 to 11,100 years BP, provides an insight into the duration of the wheat domestication process. A progressive substitution by domesticated cereals is then observed for the period between 11,000 to 10,500 years BP; and finally, the domination by the latter forms is apparent during the period from 10,500 to 7,800 years BP. The domestication phase was thus spread over several centuries, and simultaneously involved several species: einkorn (*T. monococcum*) and emmer (*T. dicoccum*) for wheat, but also barley (*Hordeum vulgare*) and rye (*Secale cereale*) for other cereals (Fuller 2007). Tanno and Willcox (2006) show a similar result at the Asikli site in Cappadocia, with domestication taking place over a very long period of time, and with many cereals in mixtures, with the wild type still being present at 50%, after 1,000 years.

9.3.2.2. *A process repeated in different places*

On the scale of domesticated plant species, the various centers of domestication (Middle East, Africa, China, Middle America and the Andean zone) are not linked to one another, but emerged independently and at different periods. As far as wheat is concerned, archaeological sites in the Iranian Zagros region (Gallego-Llorente et al. 2016) suggest that agriculture (in other words, domestication here) appeared several times in the Middle East (Anatolia, Levant, Middle Euphrates, Zagros Mountain). A comparison of the proportion of dehiscent (wild) and non-dehiscent (domesticated) chaff (mixed with soil to shape house walls), derived from charred material collected in Syria from four archaeological excavation sites, shows a slow progression of domesticated lineages in relation to wild wheat: from 10% cultivated lineages 10,000 years ago, this rate rose to 30% around 8,500 years ago and to 60% around 7,500 years ago (Tanno and Willcox 2006). This work suggests that humans did not domesticate wheat over a short period, and possibly in a single place or region, but that the wild-to-domesticated transition took place over a long period and potentially in different places.

9.3.2.3. *A process that started in the Middle East*

The first hypothesis to explain why the Middle East is the center of origin of wheat cultivation suggests that this was the region in which wild species, at different ploidy levels, were found, enabling the domestication and hybridization illustrated in Figure 9.2.

The second hypothesis is that this steppe region had the right pedo-climatic conditions for agriculture, notably light, easy-to-work soil and the absence of grasses to compete with cultivated cereals. Agriculture probably originated in the Fertile Crescent, between 12,000 to 9,500 years BP, during the Younger Dryas period (12,900–11,700 years BP); this was characterized by a worldwide drop in temperatures, as well as winter frosts and increasing scarcity of rainfall in the Middle East.

On the basis of these elements, a third hypothesis could suggest that this period of climatic constraint could have led to a reduction in available food resources and thus led, or even forced, human societies to cultivate the plant species they have been gathering until then (Salamini et al. 2002). The lack of resources, as a result of climate change and demographic pressure, is, undoubtedly, the most likely hypothesis for the origin of agriculture through the domestication of cereal species, particularly wheat, in the Middle East.

9.3.3. *Domestication traits*

9.3.3.1. *The impact of domestication on wheat biology*

Plant domestication is characterized by the acquisition of various physiological and morphological traits that vary from species to species (Doebley et al. 2006), such as non-dispersion of seeds (for a threshing harvest of a maximum number of grains not lost through natural dispersion), shortened dormancy and flowering periods, increased grain size and optimal germination, all of which define what is known as the domestication syndrome (Zohary and Hopf 1988).

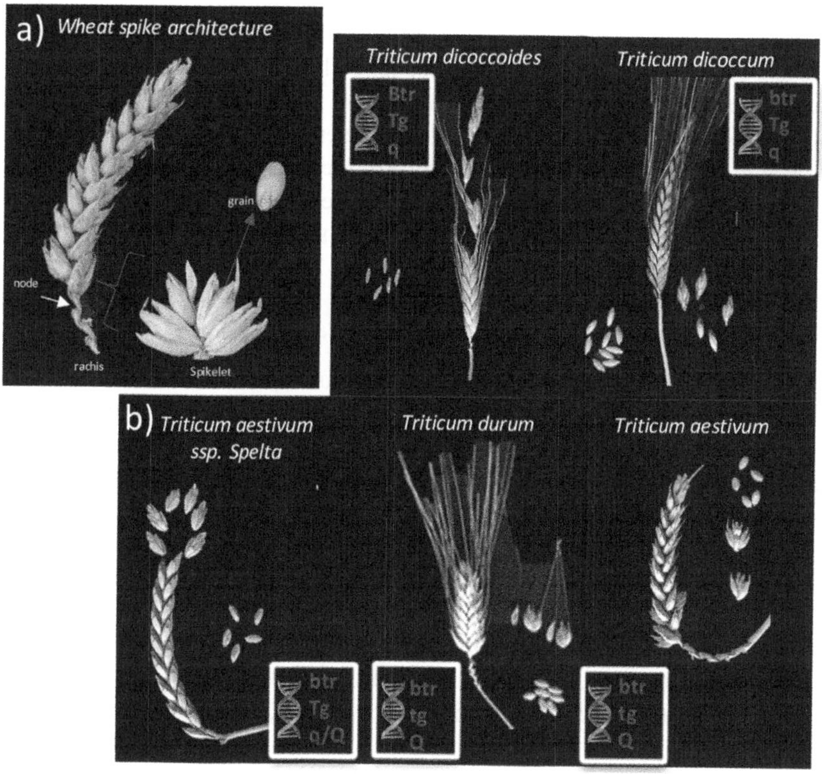

Figure 9.3. *Spike morphology in wild and domesticated wheat (source: C. Pont and J. Salse). For a color version of this figure, see www.iste.co.uk/pelletier/genetics.zip*

COMMENT ON FIGURE 9.3.– *(a) Description of wheat ear structure; (b) illustration of wild (T. dicoccoides) and domesticated tetraploid (T. dicoccum, T. durum) and hexaploid (T. aestivum) wheat ears, their morphological traits (strong or fragile rachis, naked or hulled grain) and the associated domestication alleles (Q, tg, btr): Btr: fragile rachis (ear axis); q: brittle, loose ear and hulled grain; Tg: rigid lemma and hulled grain.*

These morphological traits, which are unfavorable to the plant's development and even to the species' survival in "Nature", have not been deliberately selected by man, but are the product of natural selection, linked to the cultivation practices that have been progressively implemented. At each harvest and generation cycle, within the mixture of the species sown, these practices contribute to the selection of individuals carrying these traits, which gradually characterized the majority of the harvested plants. Indeed, annual harvesting of material has, by definition, germinated, with large-grains facilitating and accelerating the germination of grains sown deep in the soil in a short period of time of the year; this leads to the selection of lineages with more uniform flowering and ripening, and ultimately harvested through threshing, overall generalized lineages with solid, non-brittle rachis, limiting grain loss.

9.3.3.2. *The genetic basis of domestication traits*

The most visible feature is the loss of the grain dispersal mechanism once the grain has reached maturity. While wild wheat spikes break naturally (brittle ear axis) and let the spikelets fall to the ground where the grains will germinate, domesticated wheat spikes are solid. As a result, the spikes can be harvested easily before separating the grains (glumes) via threshing. This characteristic is used by archaeobotanists as a marker to identify domesticated wheat, which is distinguished by a strong rachis (ear axis) controlled by the *BTR* genes (for Brittle Rachis), and their *btr* alleles, then bearing a rough scar at the point of attachment of the spikelet that results from threshing of the ears (Zeng et al. 2020). Seed dispersal in wild *Triticeae* species is achieved by a process of disarticulation, affecting various parts of the mature inflorescence. In the "brittle rachis" types, the rachis disarticulates at several points; whereas, in the domesticated species, this occurs more along the axis of the spikelet (Figure 9.3). The site of rachis disarticulation varies between species, and in wild wheat, it occurs above a rachis node (Zeng et al. 2020). The *BTR* genes (located on chromosome 3) encode transmembrane proteins[2]

2 A cell membrane protein, whose polypeptide chain crosses the membrane one or more times in a helical conformation.

(Zeng et al. 2020), some copies of which are expressed during spike formation, with domesticated wheat with non-brittle rachis arising from sequence variations in these genes (Pourkheirandish et al. 2018). Therefore, in tetraploid wheat, the "domesticated" *btr* alleles (as opposed to wild *Btr* alleles) of the *BTR* gene carry a 2 base pair deletion for the *btr1-A* copy (introducing a premature termination codon of the ***translation***[3], truncating the functional wild protein) and a 4 kilo base pair insertion for the *btr1-B* copy, encoding a longer protein compared to the wild protein, thus inducing a loss of function of the native proteins.

Both wild and domesticated Einkorn (*T. monococcum*) and wild emmer (*T.turgidum ssp. dicoccoides*) possess *Tg* (tenacious glume) genes, carried on chromosome 2 (Faris et al. 2014), which derive strong glumes with hulled grains. These genes have undergone sequence mutations (*tg* alleles), giving rise to durum wheat (*T. turgidum ssp. durum*) and naked grain hexaploid wheat (*T. aestivum*), with soft, non-adherent glumes, facilitating large-scale grain harvesting.

The *Q* gene is a key gene in the domestication of wheat. It also controls sturdiness of the rachis, with the spontaneous mutation giving the domesticated allele (*Q*) carried by domesticated wheat, which also affects the shape of the spike (compact, solid ear), resulting in a grain that easily sheds its glume (naked wheat). The wild allele (*q*) gives a "speltoid" phenotype, characterized by extended, loose ears with an elongated rachis and an adhesive glume (hulled wheat), requiring threshing to extract the grain from its glume.

The *Q* gene on chromosome 5A encodes a member of the APETALA2 (AP2) family of ***transcription factors***. The *Q* allele has arisen through a spontaneous mutation, with the two alleles *Q* and *q* differing by a single nucleotide at the binding site of a microRNA (miRNA172). The mutation, by reducing miRNA172-dependent degradation of its messenger RNA, increases the transcriptional activity of the *Q* protein (Simons et al. 2006). This *Q* gene controls other traits linked to domestication: rachis fragility, glume shape and tenacity, spike architecture, flowering time, plant height, etc. (Zhang et al. 2011). This gene is at the origin of an original case of de-domestication, that is to say, returning to a wild phenotype of a domesticated species, known as ***feralization***. Indeed, semi-wild Tibetan wheat regained its wild trait of grain dispersion by inserting a ***transposable element*** into the *Q* gene, encoding a non-functional APETALA2 protein (Jiang et al. 2019).

---

3 Triplet of ribonucleotides with significance in the "genetic code"; in other words, the rules of equivalence linking a ribonucleotide sequence (RNA molecule) to an amino acid sequence (protein molecule).

A detailed analysis of the timing of these mutations during wheat evolution will provide a better understanding of the domestication process, as well as lead to further exploitation of the "primitive" alleles for wheat improvement. Genetic analyses indicated that the effects of the three genes are mainly additive, with the *Q* gene having the most profound effects on free-threshing, and that the combination of the three *btr-tg-Q* alleles (as opposed to *Btr-Tg-q*) are necessary to achieve a complete free-threshing phenotype, and thus a domesticated species. These results provide a better understanding of the timeline and possible pathways of wheat domestication and evolution that led to the formation of today's cultivated wheats.

9.3.3.3. *The potential role of polyploidy in domestication*

Many diploid plant species have been domesticated over the last 10,000 years in various centers of origin around the world. Polyploid wheats pre-existed domestication, and diploid wheats were domesticated (*T. monococcum*). However, even if a potential role for polyploidy in domestication remains to be established, it appears that certain major traits are the direct product of polyploidy. Grain hardness, a major trait for flour processing, is controlled by a locus called Ha (for Hardiness), located on chromosome 5 (Chantret et al. 2005). Sequencing of this genomic region between diploid, tetraploid and hexaploid wheats shows that the gene(s) encoding puroindolines, the proteins controlling this trait, are present in diploid species. These were lost in tetraploid wheat 500,000 years ago, and then reintroduced in hexaploid wheat via the *D* genome during AABB × DD hybridization, 10,000 years ago. Similarly, genes encoding reserve proteins, carried specifically by the *D* genome, could have played a role in the domestication of hexaploid (AABBDD) forms of wheat.

9.4. Genetic evidence of wheat origins: paleogenomics

9.4.1. *Modern diversity*

9.4.1.1. *The genealogy of modern wheats*

Advances in DNA sequencing technologies over the last few decades have made it possible to obtain the complete genome sequence of diploid, tetraploid and hexaploid wheat lineages (Guan et al. 2020).

In addition to whole genome sequencing, which is still too costly to be applied to numerous lineages of a single wheat species, it is possible to target the sequencing of gene exons, corresponding to the coding parts of genes, the entirety of which constitutes what is known as the exome. Comparing the genome or exome sequences of individuals from the same species enables genealogical (phylogenetic) studies to be carried out, given that two closely related individuals have similar DNA sequences.

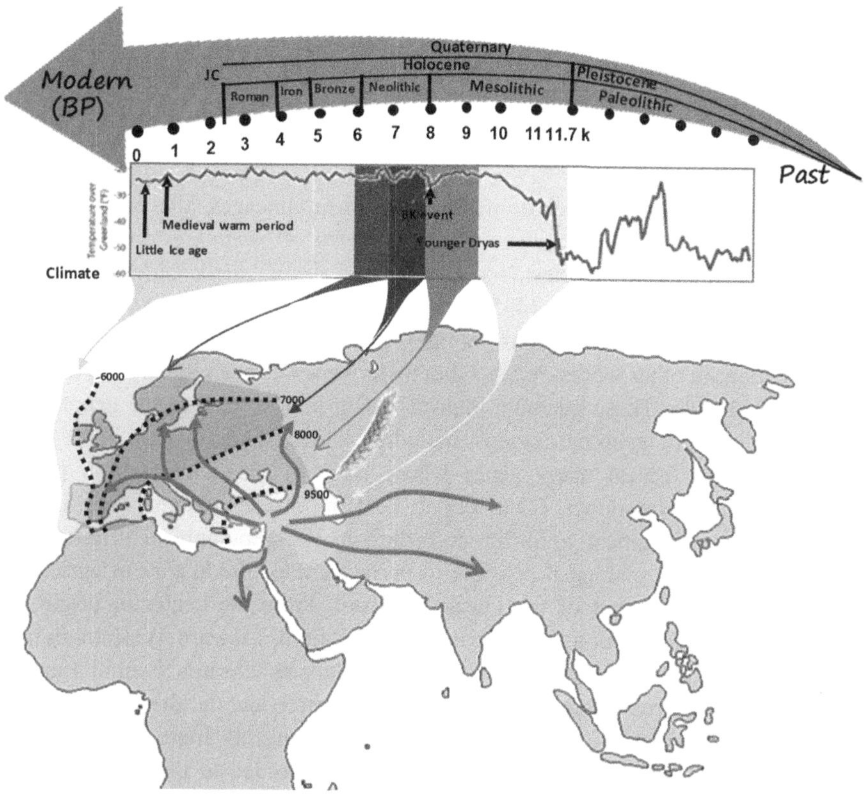

Figure 9.4. *Origins and expansion of agriculture in Europe. For a color version of this figure, see www.iste.co.uk/pelletier/genetics.zip*

COMMENT ON FIGURE 9.4.– *Geological and archaeological periods are indicated at the top (orange arrow), above their associated temperature profiles (blue curve, according to https://www.climate.gov/news-features/features/day-yesterday-when-abrupt-climate-change-came-chesapeake-bay) and major climatic changes. Known historical wheat migration routes from the site of origin (Fertile Crescent) are indicated by green arrows: westwards via an inland route (through the Balkans to Central Europe) or along a coastal route (via Anatolia to the Iberian Peninsula and the Maghreb); or eastwards via routes to the north and along the Inner Asia Mountain Corridor. The main migration phases are illustrated in different colors on the time scale (source: adapted from Pont et al. (2019b)).*

Decryption of the exome of a panel of 487 wheat lineages from 68 countries covering the world's wheat production area has produced a catalog of 620,158 ***SNPs*** in 41,032 genes. Their analysis confirms archaeobotanical data, with a phylogenetic rooting of modern wheats in the Fertile Crescent, which extends from Israel, Jordan, Lebanon, Syria, southeastern Turkey, northern Iraq and western Iran (Pont et al. 2019a).

9.4.1.2. *The genomic imprints of domestication*

Within the previous panel of nearly 500 wheat lineages, the comparison of diploid wild species (nine lineages of the *T. urartu*, *A. speltoides* and *A. tauschii* species) and tetraploid species (six lineages of the *T. dicoccoides* species) to domesticated hexaploid species (435 lineages of the *T. aestivum* species) has made it possible to identify the regions of the genome that were selected during domestication; in other words, regions that underwent a strong reduction in diversity (Pont et al. 2019a). These genomic imprints cover almost 10% of the genome and correspond to 1,221 genomic regions, including known domestication genes (*BTR*, *Tg*, *Q*), but also reveal many genes whose roles in the domestication process currently remain unknown (Gao et al. 2017). These genomic imprints of domestication also appear to differ according to the geographical origin of the lineages considered, making it possible to perfectly trace the history of agriculture and human migration out of the Fertile Crescent. From the center of origin, the spread of wheat cultivation followed two routes: firstly, toward Western Europe, through an inland route (via Anatolia and the Balkans towards Central Europe); secondly, a coastal route (via Egypt toward the Maghreb and the Iberian Peninsula); and lastly, eastwards, via routes to the north and along the Inner Asia Mountain Corridor. It was only later that other secondary events led to the colonization of African, American and Oceanic territories (Figure 9.4).

9.4.2. *Ancient DNA*

9.4.2.1. *The value of studying ancient plant DNA*

Archaeologists uncover numerous plant remains on excavation sites around the world. These fossils are often carbonized, but depending on location and context, they may be found desiccated or waterlogged from lake sediments, where traces of DNA sometimes persist over centuries and millennia. Sequencing the DNA of ancestral fossil species, at least in part, makes it possible to view the past genetic state at a given location and period of time. The study of ***ancient DNA*** requires a specific analysis, as well as experimental environments and protocols (Pont et al. 2019b). Decoding the genome of ancestors provides immediate information on the events (such as polyploidy) and mechanisms (such as sequence mutations) that have shaped the genomes of cultivated modern species. Certain adaptive variations (in gene sequence or expression, or even gene regulation), which occurred during the course of evolution, have potentially disappeared in modern species, so that they cannot be mobilized today when selecting new varieties. These genomic variations can be revealed by studying ancient DNA from archaeobotanical remains.

9.4.2.2. *Using ancient DNA to understand domestication*

Few scientific articles report on the sequencing of ancient wheat DNA, demonstrating the significance of archaeological excavations and museum collections as sources of genetic data for uncovering the history and diversity of ancient cereals. A domesticated and desiccated specimen of emmer wheat dating back to the New Kingdom (1,130–1,000 BCE) from an Egyptian museum has been sequenced and studied by R. Mott's team (Scott et al. 2019). Authenticated with typical characteristics of ancient DNA (notably the very short sequences resulting from post-mortem DNA fragmentation and the preferential degradation of cytosines, Pont et al. 2019b), the analysis used 99,078 SNPs to show that the genome of this 3,000-year-old wheat is phylogenetically related, focussing on *BTR* genes to the modern domestic emmer wheats, as well as to Indian, Omani and Turkish modern wheats, but also to wild emmer plants from the southern Levant. Another study (Wu et al. 2019) successfully shows the sequencing of ancient DNA from hexaploid wheat grains, from seven samples, which dated back to 3,200 years ago from China. This study provides important information on the origin, dispersal and genetic improvement of present-day Chinese wheat, with a new route of bread wheat dispersal and cultivation from the Qinghai-Tibet Plateau to the Yangtze River valley; this implies a crucial role of pioneer populations in the establishment of bread wheat as a staple food in the vast regions of China. Phylogenetic analyses show a close relationship between ancient wheat and today's common wheat from southwest China.

9.5. Perspectives: studying the origins and spread of wheat cultivation to support the selection of modern varieties

Archaeological and ***paleogenomic*** evidence points to the origins of wheat cultivation in the Middle East, in a region of the Fertile Crescent stretching from Israel to Jordan, Lebanon, Syria, southeastern Turkey, northern Iraq and western Iran. Humans domesticated wheat here 10,000 years ago, and since then, the species has expanded widely along an east-west axis, using four main distribution routes (Pont et al. 2019b). Over several millennia, wheat has been domesticated on the basis of naturally selected morphological traits, by harvesting lineages that have responded to cultivation practices from germination to harvest. Throughout this process, numerous varieties that were naturally adapted to the local environmental context were selected; however, these wheats sometimes disappeared, such as new glume wheat, which was relatively abundant as evidenced by numerous archaeobotanical remains, and was followed by a decline in the Bronze Age, then with the total absence from archaeological studies after the Iron Age (Kenéz et al. 2014).

It is now possible to compare ancient DNA with modern wheat diversity, as well as to study the variations in their genome sequences that have contributed to plant adaptation over the last 10,000 years, in response to hybridization (polyploidy), migration and adaptation to environmental and climatic constraints. Some of this ancestral variability is no longer present in the modern cultivated pool, and ancient DNA is now the preferred avenue of identifying this "lost" diversity. Using molecular tools such as ***genome editing***, it is now possible to study the functional role of such variations in wheat's adaptive response to environmental constraints, which have been lost during domestication or evolution, in the broadest sense. This illustrates the potential power of ancient DNA in the study of the origin, evolution and adaptation of modern cultivated plant species. Ancient plant DNA is now a privileged pathway for clarifying the genomic processes that accompanied the domestication of plants, and their adaptation, to major climatic transitions during the Holocene era. Knowledge of the genomes of these fossil ancestors provides immediate information on the events and mechanisms that shaped the genomes of modern species undergoing selection. Understanding the past adaptation of plant species can therefore open up new avenues for the selection of today's new varieties of cultivated plants, which are better adapted to current and future environmental constraints, particularly in the context of climate change.

Domestication (and, more recently, modern selection) has considerably reduced the genetic diversity of modern wheats used in breeding. First of all, not all combinations of tetraploid AABB and diploid DD wheats to form bread-making hexaploid AABBDD wheats have been domesticated-selected by humans. The possibility of recreating this type of event through hybridizations, leading to the production of new "synthetic" hexaploid wheats (AABB × DD), offers the opportunity to introduce de novo genetic and genomic diversity that has not yet been exploited in modern breeding. Over the last two centuries of selection, we have witnessed a loss of genetic diversity of over 20% between the material exploited in 1830 and that cultivated today, with variations ranging from 5.8% to 33.2%, depending on the region of the genome (Pont et al. 2019a). The loss of diversity is high (considered as selection marks) for (and in the vicinity of) genes involved in photoperiod sensitivity (*Ppd, PhotoPerioD*), the cold transition required to flower

(*VRN*, *VeRNalization*), height (*Rht*, *Reduced HeighT*), grain storage protein (Glutenins (*Glu*) and Gliadins (*Gli*)) accumulation, spike (*FZP*, *FriZzy Panicle*) and plant (*CUL*, *uniCULm*) architecture, grain number (*GNS*, *Grain NumberS*) or waxy grain (*Wx*, *Waxy*) (Pont et al. 2019a). Genetic resource centers that preserve all diversity, even old diversity, make it possible to identify material that carries unexploited genetic diversity in these global collections, which can be reintroduced into the cultivated pool through breeding programs for the development of varieties that meet the current challenges of agroecological transition and climate change.

9.6. Acknowledgments

The authors would like to thank the Agence nationale de la recherche (ANR) and the Auvergne-Rhône-Alpes (AuRA) region for funding the ArkeoAG (ANR-20-CE27-0013-01), PaleOvergne (SRESRI 2015, 23000816) and PaleoLAB (SRESRI 2015, 23001007) projects, respectively, the results of which were used to illustrate this chapter.

9.7. References

Arranz-Otaegui, A., Gonzalez Carretero, L., Ramsey, M.N., Fuller, D.Q., Richter, T. (2018). Archaeobotanical evidence reveals the origins of bread 14,400 years ago in northeastern Jordan. *Proc. Natl. Acad. Sci. U.S.A.*, 115(31), 7925–7930.

Chantret, N., Salse, J., Sabot, F., Rahman, S., Bellec, A., Laubin, B., Dubois, I., Dossat, C., Sourdille, P., Joudrier, P. et al. (2005). Molecular basis of evolutionary events that shaped the hardness locus in diploid and polyploid wheat species (*Triticum* and *Aegilops*). *Plant Cell*, 17(4), 1033–1045.

Doebley, J.F., Gaut, B.S., Smith, B.D. (2006). The molecular genetics of crop domestication. *Cell*, 127(7), 1309–1321.

El Baidouri, M., Murat, F., Veyssiere, M., Molinier, M., Flores, R., Burlot, L., Alaux, M., Quesneville, H., Pont, C., Salse, J. (2017). Reconciling the evolutionary origin of bread wheat (*Triticum aestivum*). *New Phytol.*, 213, 1477–1486.

Faris, J.D., Zhang, Z., Chao, S. (2014). Map-based analysis of the tenacious glume gene Tg-B1 of wild emmer and its role in wheat domestication. *Gene*, 542(2), 198–208.

Fuller, D.Q. (2007). Contrasting patterns in crop domestication rates: Recent archaeobotanical insights from the old world. *Ann. Bot.*, 100, 903–924.

Gallego-Llorente, M., Connell, S., Jones, E.R., Merrett, D.C., Jeon, Y., Eriksson, A., Siska, V., Gamba, C., Meiklejohn, C., Beyer, R. et al. (2016). The genetics of an early Neolithic pastoralist from the Zagros, Iran. *Sci. Rep.*, 6, 31326.

Gao, L., Zhao, G., Huang, D., Jia, J. (2017). Candidate loci involved in domestication and improvement detected by a published 90K wheat SNP array. *Sci. Rep.*, 7, 44530.

Guan, J., Garcia, D.F., Zhou, Y., Appels, R., Li, A., Mao, L. (2020). The battle to sequence the bread wheat genome: A tale of the three kingdoms. *Genomics Proteomics Bioinformatics*, 18(3), 221–229.

Hopf, M., Weiss, E., Zohary, D. (2018). *La domestication des plantes*. Actes Sud, Paris.

Jiang, Y.F., Chen, Q., Wang, Y., Guo, Z.R., Xu, B.J., Zhu, J., Zhang, Y.Z., Gong, X., Luo, C.H., Wu, W. et al. (2019). Re-acquisition of the brittle rachis trait via a transposon insertion in domestication gene Q during wheat de-domestication. *New Phytol.*, 224(2), 961–973.

Kenéz, A., Peto, A., Gyulai, F. (2014). Evidence of "new glume wheat" from the Late Neolithic (Copper Age) of south-eastern Hungary (4th millennium cal. B.C.). *Veg. Hist. Archaeobotany*, 23, 551–566.

Piperno, D.R., Weiss, E., Holst, I., Nadel, D. (2004). Processing of wild cereal grains in the Upper Palaeolithic revealed by starch grain analysis. *Nature*, 430(7000), 670–673.

Pont, C., Leroy, T., Seidel, M., Tondelli A., Duchemin, W., Armisen, D., Lang, D., Bustos-Korts, D., Goué, N., Balfourier, F. et al. (2019a). Tracing the ancestry of modern bread wheats. *Nat. Genet.*, 51(5), 905–911.

Pont, C., Wagner, S., Kremer, A., Orlando, L., Plomion, C., Salse, J. (2019b). Paleogenomics: Reconstruction of plant evolutionary trajectories from modern and ancient DNA. *Genome Biol.*, 20(1), 29.

Pourkheirandish, M., Dai, F., Sakuma, S., Kanamori, H., Distelfeld, A., Willcox, G., Kawahara, T., Matsumoto, T., Kilian, B., Komatsuda, T. (2018). On the origin of the non-brittle rachis trait of domesticated einkorn wheat. *Front. Plant Sci.*, 4(8), 2031.

Salamini, F., Ozkan, H., Brandolini, A., Schäfer-Pregl, R., Martin, W. (2002). Genetics and geography of wild cereals domestication in the Near East. *Nat. Rev. Genet.*, 3, 429–441.

Scott, M.F., Botigué, L.R., Brace, S., Stevens, C.J., Mullin, V.E., Stevenson, A., Thomas, M.G., Fuller, D.Q., Mott, R. (2019). A 3,000-year-old Egyptian emmer wheat genome reveals dispersal and domestication history. *Nat. Plants*, 5(11), 1120–1128.

Simons, K.J., Fellers, J.P., Trick, H.N., Zhang, Z., Tai, Y.S., Gill, B.S., Faris, J.D. (2006). Molecular characterization of the major wheat domestication gene Q. *Genetics*, 172, 547–555.

Tanno, K. and Willcox, G. (2006). How fast was wild wheat domesticated? *Science*, 311(5769), 1886.

Weiss, E., Kislev, M.E., Simchoni, O., Nadel, D., Tschauner, H. (2008). Plant-food preparation area on an Upper Paleolithic brush hut floor at Ohalo II, Israel. *Journal of Archaeological Science*, 35(8), 2400–2414.

Willcox, G. and Stordeur, D. (2012). Large-scale cereal processing before domestication during the tenth millennium cal BC in northern Syria. *Antiquity*, 86(331), 99–114.

Wu, X., Ding, B., Zhang, B., Feng, J., Wang, Y., Ning, C., Wu, H., Zhang, F., Zhang, Q., Li, N. et al. (2019). Phylogenetic and population structural inference from genomic ancestry maintained in present-day common wheat Chinese landraces. *Plant J.*, 99(2), 201–215.

Zeng, X., Mishina, K., Jia, J., Distelfeld, A., Maughan, P.J., Kikuchi, S., Sassa, H., Komatsuda, T. (2020). The brittle rachis trait in species belonging to the Triticeae and its controlling genes *Btr1* and *Btr2*. *Front. Plant Sci.*, 22(11), 1000.

Zhang, Z., Belcram, H., Gornicki, P., Charles, M., Just, J., Huneau, C., Magdelenat, G., Couloux, A., Samain, S., Gill, B.S. et al. (2011). Duplication and partitioning in evolution and function of homoeologous Q loci governing domestication characters in polyploid wheat. *Proc. Natl. Acad. Sci. U.S.A.*, 108(46), 18737–18742.

Zohary, D. and Hopf, M. (1988). *La domestication des plantes : origine et diffusion des plantes domestiquées en Asie du Sud-Ouest, en Europe et dans le bassin méditerranéen.* Oxford Science Publications, Oxford.

10

A History of Cultivated Rice Genomics

Philippe CUBRY, Mathias LORIEUX, François SABOT and Alain GHESQUIÈRE
DIADE, IRD, Montpellier, France

10.1. The history of rice: wild rice and cultivated rice

10.1.1. *The genus Oryza*

The genus *Oryza* comprises 23 species distributed across all continents. Based on interspecific crosses and cytogenetic studies,[1] nine ***diploid*** genomes ($2n = 24$) are recognized, and various combinations between these are observed at the ***tetraploid*** level ($2n = 48$) among 12 species. The A-genome groups the species of the *Sativa* group together, and in particular the two cultivated species, *Oryza sativa L.* and *Oryza glaberrima Steud.* The wild species *Oryza rufipogon* Griff., which is considered the origin of the Asian cultivated rice *O. sativa*, has been the subject of numerous taxonomic, ***phylogenetic*** and population studies. It can be considered a complex species with a wide diversity of biological types, ranging from perennial to annual forms, with ***autogamous*** forms, associated with a wide range of habitats, from permanently flooded areas to temporary ponds. Other studies have preferred to discuss the autogamous annual forms of South and Southeast Asia separately under the terminology *Oryza nivara*, and to reserve the name *O. rufipogon sensu stricto* for the perennial, ***allogamous***, ***photoperiod***-sensitive forms, which are widely distributed from South China and South Asia to New Guinea and Australia

1 Cytological characterization of chromosomes.

Genetics of Domestications,
coordinated by Georges PELLETIER.

(Second 1985). Lastly, separate species names have been assigned to the South American (*Oryza glumaepatula*) and Australian (*Oryza meridionalis*) forms. On the African continent, the situation of wild rices of the A-genome is much more contrasted, with *Oryza breviligulata* (*A. Chev.* and *Roehr*, equivalent to *Oryza barthii*, *A. Chev.*), an autogamous annual species that is characteristic of temporary ponds in the Sudano-Sahelian zone, and *Oryza longistaminata A. Chev* and *Roehr*, a typical species of regularly flooded areas throughout Africa and Madagascar, whose durability is expressed by the presence of vigorous rhizomes.

Various reproductive barriers exist between all these wild forms/species, of which the pollen sterility of hybrids is the most obvious manifestation. ***Gene flow*** is, nevertheless, possible and is preferentially directed from *O. nivara* to *O. rufipogon*. In *O. rufipogon*, the existence of these gene flows between its different forms, as well as with cultivated types of rice, is at the root of an entire intergradation of intermediary and adventitious[2] forms, which today only gives a biased representation of the situation pre-existing domestication and also complicates interpretations (Second 1985). Reproductive isolation is most marked with *O. longistaminata* and is characterized by albumen deterioration of the ***F1*** grains with most species, which is more or less premature, depending on the direction of crossing.

Since the 1970s, the use of ***isoenzyme markers*** has permanently clarified hypotheses on the relationships between species and domestication pathways. In particular, it confirmed the domestication of African rice *O. glaberrima* from *O. barthii*, and its independence from that of *O. sativa*. In particular, the domestication of African rice excludes the participation of the perennial forms represented by *O. longistaminata*. In Asia, the structuring of *O. rufipogon* forms is continuous between biological types but allows Chinese forms to be contrasted with those from South Asia. This observation, combined with the distinction of *O. sativa* varieties into two varietal groups, *indica* and *japonica*, long recognized on the basis of agro-morphological traits, was a strong factor in suggesting that *O. sativa* was derived from two independent domestications in South Asia and China, respectively (Second 1985).

Analysis of genomic sequence data and the dating of their divergence by ***transposable elements*** later largely confirmed that the divergence between *indica* and *japonica* (0.5–1 MA) predated domestication (Ma and Bennetzen 2004). Therefore, the very great diversity of *O. sativa*, taken as a whole, stems from a dynamic of reciprocal ***introgressions*** between ancestral varieties, which was amplified by the movement of varieties across the Asian continent (Second 1985).

2 Plants that have not been sown.

Subsequent analyses of a diverse variety of rice have clarified the intraspecific structuring of *O. sativa* as ***molecular markers*** such as ***microsatellites*** have been developed. All analyses show that bipolarization into two varietal sub-species-like groups with few intermediate situations is an essential element in the genetic structuring of *O. sativa*, even if the identification of additional groups was already possible with a small number of isoenzyme markers; it is thus possible to characterize a group related to the *indica* pool (*Aus-Boro* group), as well as a group related to *japonica* varieties (*basmati* group). Tropical rain varieties, long considered to be *indica* varieties due to the elongated shape of their kernels, have been confirmed as belonging to the tropical *japonica* sub-group. The more recent use of massive sequence data enables the eco-geographical structure of *O. sativa*, as well as the analysis of additional groups, to be continually refined; this was made possible with the set of 3,000 sequenced genomes (Wang et al. 2018) and the characterization of a subtropical group in *japonica*.

In a very different way, the variability of the African cultivated rice species *O. glaberrima* is much lower and is not structured into variety groups. However, two ecotypes of *O. glaberrima* have been recognized, with a photosensitive floating form (in other words, one whose growth allows the plant to remain above water during floods) that is characteristic of the inland delta of the Niger River in Mali, where *O. glaberrima* has been domesticated, and an earlier, upright rainfed form adapted to slash-and-burn farming in the forest zone. The cultivation of *O. glaberrima* was already considered to be in decline by botanists at the beginning of the 20th century, in view of the spread of *O. sativa* varieties introduced by the Portuguese in the 16th century (Portères 1962; Cubry et al. 2018).

The two cultivated rice species are separated by a very strong reproductive barrier. Reproductive isolation between the two species, given that they have been in each other's presence, has been particularly effective, as evidenced by the absence of reciprocal introgressions. Despite this, it is possible to overcome this reproductive barrier through backcrossing, making it possible to use this species to introduce genes of interest in Asian rice, which creates interspecific varieties such as NERICA (*NEw RICe* for Africa), popularized by AfricaRice since the 2000s (resistant to numerous diseases, drought and high night-time temperatures).

In the 1980s, several non-exclusive hypotheses were already put forward to explain the limited success of *O. glaberrima*, such as a lack of diversity in its wild progenitor, its genetic isolation from other species in Africa, the much less diverse environments than in Asia, or the rural African communities' preference for other cereals, such as millet and sorghum.

10.1.2. *Characteristics of cultivated rice*

Compared to the wild species from which they originate, cultivated plants are distinguished by traits resulting from domestication. Among these, the loss of shattering (shattering being the ability of mature grains to detach from their parent plant) is the most common domestication trait to unambiguously reveal a cultivated form in grasses. Therefore, the presence of grains still attached to the rachis of panicles (rice inflorescence) is decisive in confirming the cultivated nature of archaeological remains and providing arguments for domestication. Nevertheless, shattering is a complex trait that depends on several genes with a major effect, such as *sh4*, or a partial effect (***QTL***), such as *qSH1* and *qSH3* in the case of Asian rice. They interact and can give rise to intermediate phenotypes (Ishikawa et al. 2017), as in some current rice varieties (varieties of *indica* and basmati rice) that are characterized by high sensitivity to shattering. Distant crosses between genetically distant varieties can also give rise to curved forms (such as a seed ending in a ridge) with varying degrees of shattering, reflecting the greater genetic diversity behind these phenotypes (weed rice, red rice, Camargue *crodo* rice). More upright phenotypes (*Prog1*) with earlier, more uniform flowering (*Hd3a*, *Hd8*, *Hd2*), linked to more annual forms, are likely to have accompanied the early stages of rice domestication (Hu et al. 2018).

The photoperiod sensitivity characteristic of *O. rufipogon* has not, however, been completely lost in Asian rice; it has the important role of setting the crop cycle of traditional varieties grown in wetlands, without irrigation control. Inversely, the near loss of any reaction to the photoperiod, combined with powerful root development, is an example of newly acquired traits that have enabled adaptation to the strict rain conditions that first developed in Southeast Asia and then, with the introduction of *O. sativa*, in Africa.

The most obvious manifestation of domestication can be seen in the complexification of panicle structure (part of the plant bearing the flowers, then the seeds, composed of a central axis and more or less branched secondary axes), with the increase in the number of first-order branches and the appearance of secondary branches, which have considerably increased the number of grains produced by each panicle. The domestication of Asian rice was also accompanied by exceptional diversity in grain size and shape, including new aspects linked to the organoleptic value of the seeds (amylose content, vitreosity, aroma, etc.). The results of very high-density genetic mapping[3] of the majority of traits distinguishing *O. rufipogon* from cultivated rice have enabled us to identify 58 QTLs, mostly with weak to

3 Relative positioning of genes on different chromosomes.

moderate effects; 10 of these very specifically correspond to previously identified genes.

10.2. The beginnings of genomics and the pan-genomic revolution

10.2.1. *The first genomic sequence*

In September 1997, the International Rice Genome Sequencing Project (IRGSP) was launched, involving teams from all over the world (Japan, France, the United States, South Korea, India and China). The project aimed to obtain the genome sequence of the Nipponbare variety of *O. sativa ssp japonica*. Another independent project was based on the 93-11 variety and was supported by the Chinese Superhybrid Rice Genome Project (CSRGP). The two genomes of Nipponbare and 93-11, each representing one of the *O. sativa* subspecies (*japonica* and *indica*, respectively), was made available for the first time in 2002. The sequences obtained, which are still fragmented and partial, cover around 95% of the total genome sequence of each of these varieties. Numerous efforts were made over the following years to finalize the Nipponbare sequence, culminating in the so-called IRGSP 1.0 version in 2013, which covered over 99% of the Nipponbare genome.

10.2.2. *Complementary sequencing and 3,000 genomes*

In the years 2005–2015, following the arrival of so-called second-generation sequencing technologies, numerous other rice sequencing projects were launched. These sequences were not available with the level of quality of the Nipponbare or 93-11 genomes (whole or nearly whole chromosomes) but were as small fragments or simply as variations around the Nipponbare genome. This approach is commonly referred to as resequencing. Therefore, in 2012, Bin Han's team published sequence data for 1,083 cultivated and 446 wild *O. rufipogon* rice ***accessions*** (Huang et al. 2012), allowing a new vision of rice domestication to be proposed (see section 10.3). The culmination of these resequencing efforts was the publication of more than 3,000 genomes from accessions covering all (or most) of the genetic diversity of *O. sativa*. These public data, the result of an international collaboration led by IRRI (International Rice Research Institute, Philippines), CAAS (Chinese Academy of Agricultural Science) and BGI (Beijing Genomic Institute), have since made it possible to identify numerous variations in the Nipponbare or 93-11 genes between rice varieties, as well as uncovering the limitations of these two reference genomes.

10.2.3. *New reference genomes and the appearance of rice pan-genomes*

In the course of several studies of various genes of agronomic interest, it became apparent that certain genes were not present in either the Nipponbare or 93-11 genomes. The two best-known examples are *Sub1A*, which provides resistance to submergence (when the aerial part of the plant is below the surface of the water), and *Pstol1*, which provides tolerance to phosphorus deficiency in soils. Scientists therefore began sequencing other varieties in order to identify genomic, rather than genetic, differences, and thus find new genes.

As a result, several teams produced new, almost complete "reference" genomes, mainly based on the third-generation long-read sequencing (such as the genome of the Kasalath variety). This was followed by the publication of several complete genomes with very useful resources, such as the 12 very high-quality genomes produced from varieties representing the structure of the species, published by Rod Wing's team (Arizona Genomics Institute, KAUST).

These new genomes, and especially their comparison with existing genomes, highlight the need for a ***pan-genomic*** approach to rice diversity (see Box 10.1).

One of the first analyses of this type by D. McCombie's team (Cold Spring Harbor Laboratory) focused on three varieties and revealed different "gene spaces" between these three groups: between 300 and 800 genes specific to each individual, and between 471 and 622 genes that are only common to two (for 37,200 genes common to all three genomes). At the same time, it was determined from the sequencing of 1,483 cultivated rice varieties that the rice genome is made up of over 60% distributed sequences (in other words, absent in at least one individual of the species) and 40% of the agronomic traits of interest are more or less linked to these distributed sequences. Re-analysis of the 3KRGP data (Wang et al. 2018) from this perspective thus shows that over 50% of the genome of these varieties is distributed and identifies 12,465 new genes that could not have been detected by only considering the first reference genomes. More recently, Qin et al. (2021) have even shown that an additional 1.3 Gb can be identified in comparison with the Nipponbare reference genome – which only comprises 390 Mb – with 33 fully assembled cultivated rice genomes (chromosomes).

New rice genomics studies are thus moving toward pan-genomics, which also makes it possible to reassess certain aspects of domestication (see sections 10.3.1–10.3.3).

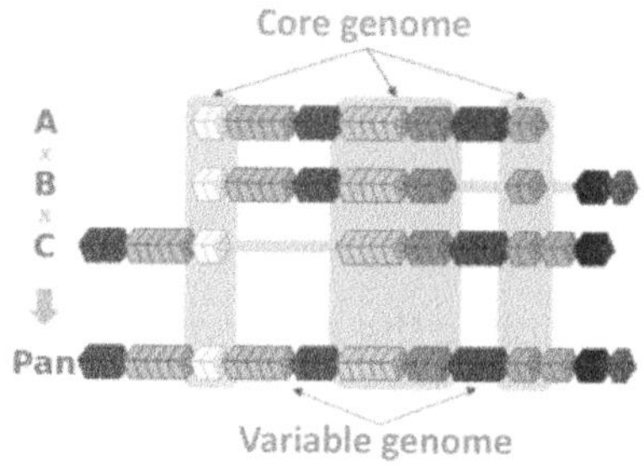

Figure 10.1. *Diagram illustrating the concepts of pan-genome, core genome and distributed (variable) genome for a group of three individuals A, B and C (source: Durant et al. 2019). For a color version of this figure, see www.iste.co.uk/pelletier/genetics.zip*

COMMENT ON FIGURE 10.1.– *For each individual, groups of genes are represented by different colors. The core genome groups together genes that are present in all individuals; the distributed genome, which are present in some individuals; and the pan-genome, which are genes present in all three individuals (approved by E. Durant).*

The notion of a distributed genome is still hotly debated, both in terms of its connotation (sequences in the distributed genome may, in fact, have an important function) and its grain (some authors distinguish between sequences with a high presence – or shell – and those with a low presence – or cloud). The distributed part of a *clade* can be much larger than the genome size of an individual, and thus the core part. For example, in the grass *Brachypodium distachyon*, the *pan-genome* has a minimum size of 430 Mb; whereas, the genome of a single individual is only 230 Mb. In other words, looking at a single individual as the basis of diversity leaves out a huge amount of genomic resources, and therefore genes, that are available for both adaptation and varietal improvement.

Box 10.1. *The pan-genome concept*

10.2.4. *Other rice species*

The development of second and third-generation sequencing technologies has made it possible to sequence at least one genome representing all, or almost all, species of the genus *Oryza*, and at different levels of completion and quality. The international iOMAP project, led by R. Wing (University of Arizona and KAUST), aims to produce reference genome sequences for all species in the genus. With regard to cultivated rice, in addition to the various sequences for *O. sativa* and its ancestor *O. rufipogon*, an effort has been made on the African cultivated rice

O. glaberrima, with a first version of the genome of the CG14 variety produced by the iOMAP consortium in 2014. Shortly afterward, the sequencing and assembly of three new *O. glaberrima* genomes (*CG14*, *TOG5681* and *Og6*; Monat et al. 2017) and their comparison showed significant pan-genomic diversity.

10.3. Genomics' contribution to the study of rice domestication

Most of the traits linked to the domestication process were selected during the initial phases of domestication; others in a more recent post-domestication phase by cultivators. This selection has left its mark on the genomes of cultivated plants, and can therefore be revealed using modern sequencing and genome composition analysis tools. Several methods can be used to study the genomic basis of domestication and related traits, in particular:

– genetic mapping or association studies, aimed at identifying and characterizing the *loci* (genes, QTL) involved in variations in traits of interest;

– "genome scan", in other words, genome-wide analysis that can be carried out in successive windows of a few loci in order to identify "atypical" allelic variants that are likely to have a particular evolutionary and selective history.

In rice, numerous studies have identified or confirmed genes that are involved in variations in the traits listed above. This is the case, for example, of the *prog1* gene, involved in the upright position of plants in both cultivated rice species, or of a number of genes linked to flowering or dehiscence.

10.3.1. *Genomics as a tool for studying domestication traits*

Before the advent of genomics, the search for genes involved in domestication traits required the creation of populations that were derived from controlled crosses between domesticated and non-domesticated rice varieties (or those returned to a similar state, such as weed or red rice). These populations were then characterized using molecular markers, and the presence of traits considered specific to domesticated plants was sought. The search for statistical associations between these traits and certain molecular markers made it possible to identify loci that were involved in domestication (Bres-Patry et al. 2001).

This approach, known as genetic mapping, is a very powerful way of identifying major effect genes (where variation in a single gene explains variation in a trait, which is then said to be "monogenic"). In the case of multigenic traits, whose variation generally appears to be continuous, such as height, which depends on several partial-effect genes (in this case, referred to as QTL), this approach lacks precision, notably due to the limited number of recombinations in these populations at ***meiosis***.

The possibility of obtaining massive genomic diversity data for large populations has revolutionized gene and QTL mapping: ***genome-wide association studies (GWAS)*** can now be conducted to search for significant associations between genetic variants (mostly ***SNPs***) at different positions in the genome with variations in agronomic traits of interest. While this type of approach is similar to more conventional genetic mapping methodologies, the much higher resolution of these association studies is a result of the presence of ancestral recombination points over many generations. Their use on populations not derived from controlled crosses has led to real advances in our understanding of the genetic bases of quantitative variations in traits that may be linked to domestication. These studies have enabled us to identify new sources of functional diversity for traits linked to flowering and drought.

More recently, with the rise of ecological or landscape genomics, these association approaches have been extended to the study of the impact of environment and climate on genomes and, through this, to the understanding of local adaptation. These approaches make it possible to identify loci showing allelic variations, linked to those of key climatic variables for the plant. A recent example in African rice is the identification of candidate genomic regions for temperature adaptation, which are interesting sources of variability for the improvement of varieties in the face of climate change. Nevertheless, these studies, although potentially very powerful, require experimental validation through functional analysis of the candidate loci identified. While the link to the history of domestication may not immediately appear to be obvious, these approaches enable us to identify variations selected according to environments, particularly in traditional varieties, which may have a specific history that diverges from that of the majority of the genome. This is the case, for example, for variations that may have been acquired post-domestication in a given environment from local wild relatives that may have different genetic bases from the wild relatives of the initial domestication zone. This raises the question of the importance of possible gene flow from wild to cultivated forms during late phases of domestication, with the inclusion of wild sequences providing an adaptive advantage.

As already mentioned, the search for ***selection signatures*** on a genome scale involves identifying unusual variants, whose frequency in populations significantly deviates from the majority of other variants. This type of approach, which identifies candidate markers linked to selection (whether they be natural or artificial), can partly overlap with studies of association with climatic variables in the history of domestication. We could, for example, look at the evolution, along the genome, of differences in diversity between wild and cultivated populations. The aim of such an approach would be to identify areas where the loss or gain in diversity is greater than the genome average, thus signaling areas of particular significance for domestication.

10.3.2. *Genomics as a tool for a better understanding of domestication history*

As mentioned above, two independent domestications led to the cultivation of rice in Asia and Africa. Recently, archaeological evidence has also revealed a third domestication, in South America, of a rice that has since disappeared. Each of these domestication events has its own characteristics, and certain scenarios established on the basis of archaeological data are clearly competing. Recent genomic studies are shedding some light on these histories and allowing us to propose, confirm or refute hypotheses on the origin of cultivated species, and move away from a paradigm in which we attempt to reconstruct a history of domestication that would be consistent with piecemeal evidence, and instead toward a genuine hypothesis test based on the characteristics of genomes and the models and tools of population genomics. The questions that can be addressed by such approaches cover the major issues of the location, temporality and modalities of domestication.

One of the first questions that can be answered by these approaches is that of the temporality of domestication. Indeed, one of the most widely accepted postulates is that a major event such as bringing an organism into cultivation must have had an impact on its demographic, notably by creating what is known as a genetic bottleneck; that is to say, a significant reduction in diversity and a loss of *alleles* linked to a low population size. However, recent approaches initially developed in humans, using tools derived from ***coalescent analysis***, make it possible to take advantage of the vast amount of information that is present in genomes to reconstruct past demographic history. If we can identify and date a loss of diversity, such as that expected at the time of domestication, we can then validate or put forward hypotheses about this domestication. This is what researchers have achieved for the study of Asian rice (Civáň and Brown 2018), as well as African rice (Cubry et al. 2018) (see sections 10.3.3 and 10.3.4).

Another possibility put forward by the concepts and methods of population genomics is the modeling of the spread of a cultivated species from its area of origin, and thus the possibility of statistically determining this area of origin. For example, genomic datasets can be simulated under the assumption of geographic spread. The simulated datasets are then compared with the real dataset in order to identify which are compatible, and thus obtain information on the probable origin of domestication. This type of approach has been used for both Asian and African rice.

Lastly, we can study the genomic variants selected by the domestication process. To do this, we can mobilize the approaches and techniques mentioned above. These analyses can shed light on the genes and molecular mechanisms involved in domestication.

10.3.3. *Domestication of Asian rice*

The study of the genetic diversity of *O. rufipogon* gained new traction in the 2000s with the work of Chinese laboratories (Beijing Institute of Genomics, Shanghai Institute of Biological Sciences), which developed the first large-scale genomic studies and simultaneously made it possible to integrate numerous samples of Chinese origin that had been lacking in previous analyses.

One of the leading studies in the field is based on the sequencing of over 400 *O. rufipogon* accessions and 1,000 *O. sativa* varieties (Huang et al. 2012). It confirms a much greater diversity of *O. rufipogon* compared with *O. sativa*, but with little differentiation, even if three major groups can be identified, one of which, in southern China, has a strong affinity with *japonica* varieties. The situation is less clear-cut in India, leading these authors to prefer a model in which primitive domestication first took place in China, with varieties then being hybridized with *O. rufipogon* strains from South and Southeast Asia to give *indica* varieties and other varietal groups. Largely using the same genomic data, but inferring the diversity patterns of *O. rufipogon* ancestral populations from the diversity of current varieties, other authors position the ancestors of cultivated rice differently, and even suggest a separate domestication of *Aus-Boro* varieties. By studying the history of gene flow between two groups of varieties, these same authors conclude that reciprocal exchanges between *indica* and *japonica* varieties have been limited, while at the same time highlighting the complexity of *aus* varieties (Civáň and Brown 2018). Therefore, genomic studies may lead to contradictory conclusions, but do not necessarily univocally exclude one or the other scenarios (independent introgressions/domestications) that may have occurred throughout Asia, and throughout the duration of domestication. Lastly, failure to consider the ecotypes of *O. rufipogon* samples and the importance of gene flows from crops to wild rice can strongly bias the evolutionary interpretations of genomic analyses. Indeed, considering *feral* accessions, or accessions that have strongly integrated genes from cultivated varieties, as belonging to the wild group can lead to wrong hypotheses.

10.3.4. *Domestication of African rice*

The domestication of plants in West Africa has long been the subject of debate, with advocates of widespread domestication throughout the Sahelian zone (the non-centric domestication hypothesis) versus those of more localized domestication, for example, on the scale of the Inner Niger delta (the centric hypothesis). Recent genomic analyses, carried out on various cereals and cultivated plants of African origin, have cleared up some of these questions, particularly with regard to *O. glaberrima* rice. A reconstruction of the demographic history of the cultivated species *O. glaberrima* and the related wild species *O. barthii* has revealed a sharp demographic decline in *O. barthii* and the gene pool that gave rise to the *O. glaberrima* species, which began more than 10,000 years ago; this was probably linked to the aridification of the Sahara. Afterward, the two species appear to follow different demographic trajectories, with the wild species undergoing rapid expansion around 3,500 years ago, while the cultivated species underwent a demographic explosion just over 2,000 years ago. This period (3,500–2,000 years BP), when demographic histories diverge, has been suggested as the start of African rice domestication (Cubry et al. 2018). The same authors also looked at the location of this domestication. Using a model incorporating the impact of the spatial component in genetic simulations, they were able to demonstrate that the most likely area of origin for African rice is that of the inland delta of the Niger River, in present-day Mali. Interestingly, these results perfectly corroborate those put forward on the basis of archaeological or ethnolinguistic data (Portères 1962), with the oldest archaeological remains of cultivated African rice grains having been found in this geographical region.

African rice has a much shorter and simpler domestication history than Asian rice, and its cultivation has remained largely confined to the African continent (although some populations have been described in South America and the Caribbean). Its cultivated traits also appear to be imperfect or incomplete. This is the case, for example, with loss of seed-shattering, which varies from one population to another and is generally incomplete, or with grain color traits that have not yet been fixed in certain traditional varieties. Ultimately, while the history described using the whole genome seems generally consistent with archaeological and ethnolinguistic hypotheses, it should be noted that it does not exclude more complex histories, which the power of current genomic datasets does not yet allow us to infer. For example, we can look at the geographical history and spread of key genes, and try to identify their geographical area of origin, which may be different from the area of domestication of the species. A recent study of African rice showed that key alleles of genes selected during domestication may have varied geographical origins. This raises the question of whether there was a single center of domestication, or whether domestication took place in several distinct geographical locations. It should be noted, however, that while these alleles may have originated in geographical regions other than the assumed center of domestication, they may also have been acquired through gene flow and hybridization after domestication.

10.4. Conclusion: the "continuity" of domestication

One way of clarifying certain open questions, such as gene flow between the wild and cultivated groups, would be to use archaeological samples with ***ancient DNA*** sequencing methodologies, which would enable us to better trace genomic evolutions throughout the initial domestication process. This type of genomic study on ancient DNA could also be used to study the likely divergent history of a third domestication of rice, recently highlighted in archaeological remains from South America. Such studies would be particularly relevant to understanding the factors that enabled rice to be domesticated, not just in one part of the world, but on three different continents. In addition, they would shed valuable light on the convergence between different independent domestication events, a convergence that is already highlighted for some traits between Asian and African rice, where it has been shown that certain genes and metabolic pathways were commonly affected by domestication, although the selected allelic variants are potentially different between the two species. Such studies could be carried out for key traits, such as seed-shattering and flowering. Beyond the significance of this convergent selection for domestication traits for at least two independent domestications within the same genus, these approaches could have much broader impacts, extending to an understanding of what makes a species capable of being domesticated. Identifying the genomic resources that can be used for domestication could be beneficial on a much wider scale than the genus *Oryza* and could contribute to the necessary adaptation of agriculture to the global changes we are experiencing.

In the current context of climate change, the adaptation of our domesticated plants is indeed a crucial issue. A better understanding of how domestication works, and what has made rice domestication so cosmopolitan and effective, could help to fuel the possibilities for improving cultivation, and even contribute to current thinking on neo-domestication. These approaches seek to identify plants of interest in the wild group and select (or induce) candidate variants in them for domestication. As mentioned in the Introduction, the genus *Oryza* has a large number of species, particularly tetraploids with high growth potential, which could be used in such processes. Directed mutagenesis techniques, which are emerging in rice, could be developed on certain previously identified key target genes in Asian and African rice domestication studies, with a more experimental approach. These genomic rewriting (or editing) approaches now make it possible to attempt neo-domestication by directly targeting other wild rice species (see Chapter 13).

Despite recent technological advances and the contribution of genomics, there are still many areas of uncertainty concerning the history of rice domestication. This is particularly true of the wild Asian gene pool. This appears to have been heavily hybridized by gene flow with *O. sativa*, making it difficult to use for domestication studies, and leading to erroneous conclusions about the origin of domestication. In parallel with ancient DNA approaches, another way of answering these questions could be to reconstitute "pure" ancestral gene pools from current genomes; in other words, free from introgressions of domesticated genomes, similar to the approaches carried out in humans, but which have also begun in *O. sativa* rice.

Ultimately, a better understanding and knowledge of the structures and evolutions of wild and cultivated pan-genomes, and of the forces affecting them, would enable us to better understand the mechanics of domestication. These areas of research are undergoing intensive development, not least because of the challenges involved in accumulating ever larger and more complex datasets, and in analyzing, visualizing and exploiting them.

10.5. References

Bres-Patry, C., Lorieux, M., Clément, G., Bangratz, M., Ghesquière, A. (2001). Heredity and genetic mapping of domestication-related traits in a temperate japonica weedy rice. *Theoretical and Applied Genetics*, 102(1), 118–126.

Civáň, P. and Brown, T.A. (2018). Role of genetic introgression during the evolution of cultivated rice (*Oryza sativa* L.). *BMC Evolutionary Biology*, 18(1), 57.

Cubry, P., Tranchant-Dubreuil, C., Thuillet, A.C., Monat, C., Ndjiondjop, M.N., Labadie, K., Cruaud, C., Engelen, S., Scarcelli, N., Rhoné, B. et al. (2018). The rise and fall of African rice cultivation revealed by analysis of 246 new genomes. *Current Biology*, 28(14), 2274–2282.e6.

Durant, E., Sabot, F., Rouard, M. (2019). Panache: A visualization tool for the exploration of plant pangenomes. In *5th Conference on Plant Genome Evolution*, Sitges.

Hu, M., Lv, S., Wu, W., Fu, Y., Liu, F., Wang, B., Li, W., Gu, P., Cai, H., Sun, C. et al. (2018). The domestication of plant architecture in African rice. *The Plant Journal*, 94(4), 661–669.

Huang, X., Kurata, N., Wei, X., Wang, Z.-X., Wang, A., Zhao, Q., Zhao, Y., Liu, K., Lu, H., Li, W. et al. (2012). A map of rice genome variation reveals the origin of cultivated rice. *Nature*, 490, 497–501. doi: 10.1038/nature11532.

Ishikawa, R., Nishimura, A., Htun, T.M., Nishioka, R., Oka, Y., Tsujimura, Y., Inoue, C., Ishii, T. (2017). Estimation of loci involved in non-shattering of seeds in early rice domestication. *Genetica*, 145(2), 201–207.

Ma, J. and Bennetzen, J.L. (2004). Rapid recent growth and divergence of rice nuclear genomes. *Proceedings of the National Academy of Sciences*, 101(34), 12404–12410.

Monat, C., Pera, B., Ndjiondjop, M.N., Sow, M., Tranchant-Dubreuil, C., Bastianelli, L., Ghesquière, A., Sabot, F. (2017). De novo assemblies of three *Oryza glaberrima* accessions provide first insights about pan-genome of African rices. *Genome Biology and Evolution*, 9(1), 1–6.

Portères, R. (1962). Berceaux agricoles primaires sur le continent Africain. *The Journal of African History*, 3(2), 195–210.

Qin, P., Lu, H., Du, H., Wang, H., Chen, W., Chen, Z., Li, S. (2021). Pan-genome analysis of 33 genetically diverse rice accessions reveals hidden genomic variations. *Cell*, 184(13), 3542–3558.

Second, G. (1985). Evolutionary relationships in the *Sativa* group of *Oryza* based on isozyme data. *Genet Sel Evol.*, 17, 89–114.

Wang, W., Mauleon, R., Hu, Z., Chebotarov, D., Tai, S., Wu, Z., Li, M., Zheng, T., Fuentes, R.R., Zhang, F. et al. (2018). Genomic variation in 3,010 diverse accessions of Asian cultivated rice. *Nature*, 557, 43–49. doi: 10.1038/s41586-018-0063-9.

11

Grapevine Domestication and Selection

Patrice THIS[1,2], Thierry LACOMBE[1,2] and Cécile MARCHAL[3]
[1]*UMR AGAP Institut, CIRAD, INRAE, Institut Agro, Université de Montpellier, France*
[2]*UMT Geno-Vigne®, IFV-INRAE-Institut Agro, Montpellier, France*
[3]*Unité expérimentale du Domaine de Vassal, INRAE, Marseillan, France*

11.1. Introduction

Grapevines are grown on all five continents. The global surface area of vineyards in 2021 was 7.32 million ha (International Organization of Vine and Wine[1] (OIV) 2022). The three main products from vines are, in order of magnitude, wine 48.8% (261 million hl in 2021 from 33.9 million tons of grapes), table grapes, consumed as fresh fruit 43.3% (30 million tons in 2021), and dried grapes from a dehydration process 7.9% (1.4 million tons in 2021). China is the world's leading grape producer, with 20% of global production, mainly table grapes (73% in 2021, OIV 2022). Europe is the world's leading wine producer, with production of 153.7 million hl in 2021 (OIV 2022). Italy, France and Spain accounted for 47% of world production in 2021. France, with its 798,000 ha yielding an average of around 50 million hectoliters, produced just 37.6 million hl of wine in 2021, which was down sharply on previous years due to unfavorable weather conditions, such as the late frosts in April. As a result, wine production fluctuates, as it is subject to weather hazards.

Vitis vinifera L. is the main species used to produce these grapes and wine, although a few other species of the genus *Vitis*, such as *Vitis labrusca*, are also used.

1 Available at: https://www.oiv.int/.

Genetics of Domestications,
coordinated by Georges PELLETIER. © ISTE Ltd 2024.

To this day, and since the arrival of phylloxera at the end of the 19th century, vines can only be grown grafted onto a rootstock that is resistant to this aphid of American origin. Rootstocks can sometimes come from pure *Vitis* species, such as "Rupestris du Lot"; however, more often, they are interspecific hybrids, which are often complex, usually involving two to three American *Vitis* species (mainly *Vitis rupestris*, *Vitis riparia* and *Vitis berlandieri*), and sometimes *Vitis vinifera*. Rootstocks will not be discussed in this chapter.

A key element in the wine industry is the variety, also known as cultivar.

All table grape varieties are marketed under a varietal name, the best known in France probably being "Chasselas", "Muscat de Hambourg" or "Italia". Likewise, the grape varieties used to make a wine give it its distinctive character, whether it is a varietal wine (made from a single grape variety) or a blended wine. In PDO (protected designation of origin) and PGI (protected geographical indication) zones, the number, identity and proportion of grape varieties used in wines are regulated and defined in the particular specifications for each product. There is considerable genetic and morphological diversity between grape varieties, but also within the same variety. While there are around 6,000 grape varieties in the world, only a few dozen account for nearly a quarter of the world's vineyards.

In this chapter, we first present the species *Vitis vinifera*, its botanical, genetic and agronomic particularities, as well as its great genetic diversity. We will then look at the origins of this species, its domestication from the wild ancestor *Vitis vinifera subsp. sylvestris* and its evolution to the present day. There will be particular mention of recent work in molecular genetics and genomics, which has enabled us to trace the history of this species. We will focus on two important characteristics: the sexual type of the flower, with its evolution from dioecia (from Greek *dis* "two" and *oekia* "house"), meaning that male and female flowers are borne by different plants, to hermaphroditism, and the color of the berry skin, resulting from the presence of anthocyanins.

11.2. *Vitis vinifera L.*, the main species of the genus *Vitis* used for the production of table grapes and wine

All species used for grape production belong to the family *Vitaceae* and the genus *Vitis* (Figure 11.1). Within this genus, *Vitis vinifera L.* is the main species that is cultivated worldwide. Other genera are also used to some extent, but only as ornamental plants, such as *Parthenocissus* (the "Virginia creeper", *Parthenocissus quinquefolia* or *Parthenocissus tricuspidata*), *Rhoicissus* and *Cyphostema*. Species of the genus *Vitis* are grouped into two subgenera: *Muscadinia* and *Vitis*. They are much more numerous in the subgenus *Vitis*, which comprises some 80 species native to the American and Asian continents. *Vitis vinifera L.* is the only species that is native to Eurasia. It can be divided into two subspecies: the wild form, *Vitis vinifera subsp. sylvestris* (or *silvestris*), also known as "lambrusques" (referred to as *V. sylvestris* or "lambrusques" in the rest of this document), which represents the ancestor of cultivated vine, and *Vitis vinifera subsp. vinifera* (or *sativa*), which represents the cultivated compartments (referred to as *V. vinifera* in the rest of this document).

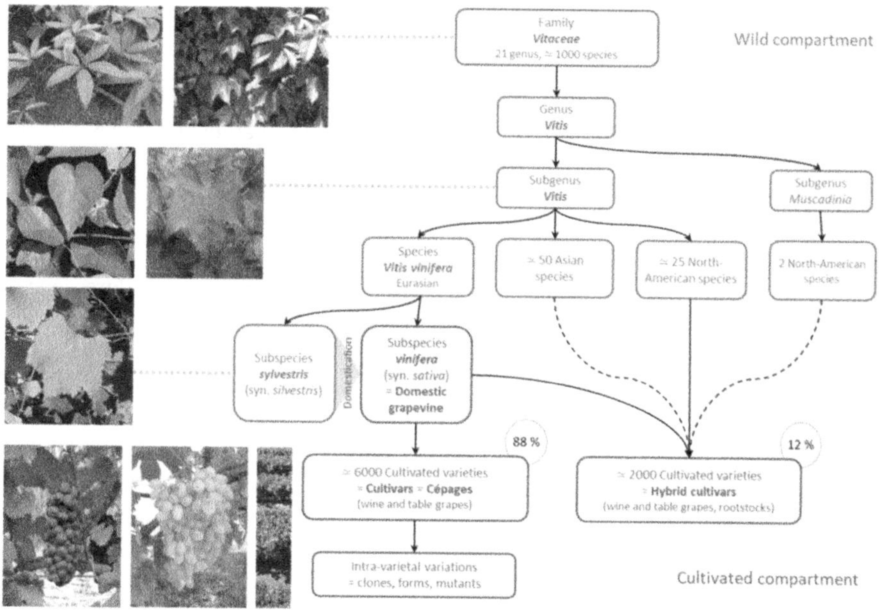

Figure 11.1. *Presentation of the vine's taxonomic position (source: INRAe illustrations). For a color version of this figure, see www.iste.co.uk/pelletier/genetics.zip*

11.2.1. *Vitis vinifera L.*: biology and genetic diversity

11.2.1.1. *The grapevine, a grafted woody perennial plant*

The grapevine is a perennial plant, with a seed-to-seed cycle of around 4–5 years. A grapevine plant can live for several decades, but winegrowing practices, as well as the biotic and abiotic stresses to which plants are subjected mean that, on average, a vineyard has to be renewed every 25–30 years. Heterozygous grapevines cannot be propagated identically by seed, as this would result in individuals very different from their parents. It is therefore generally propagated by cuttings, layering or grafting, with these methods of propagation enabling the characteristics of the varieties to be almost totally preserved (the so-called "clonal" plant).

In its natural state, the grapevine is a liana that climbs trees thanks to its long branches and tendrils at the nodes. Humans have shaped the grapevine into what we see today in vineyards around the world. The woody stems represent the shoots, also known as branches. Up until the end of the 19th century, vines were grown on their own roots. With the arrival of phylloxera in Europe around 1860 – an insect that attacks vine roots and causes the death of highly susceptible *Vitis vinifera* plants – grafting onto aphid-resistant rootstocks of American origin saved European vineyards. Today, almost all vines are grafted, making each grapevine stock a blend of two different partners: the rootstock, located in the soil and forming the roots, and the scion, the aerial part of the stock that produces the fleshy fruit or berries.

The small, green flowers bear a pistil with an ovary containing four ovules (giving the seeds) and five stamens. These are grouped together in inflorescences which, after flowering, fertilization and berry development, form the grapes.

The vast majority of *Vitis vinifera* flowers are hermaphroditic (male and female parts present in the same flower), although a few female varieties still exist. In other wild species of the genus *Vitis*, including *V. sylvestris* and the rootstocks derived from it, individuals are dioecious; in other words, there are male individuals (with male flowers that only have stamens) and female individuals (with flowers where only the female part is functional).

11.2.1.2. *A diploid plant with extreme diversity*

Apart from a few rare spontaneous mutants, all species of the subgenus *Vitis* are ***diploid***, with 19 pairs of very compact chromosomes. Species of the subgenus *Muscadinia* have 20 pairs of chromosomes.

All species of the genus *Vitis* are potentially interfertile: a few spontaneous hybrids exist in nature between certain ***sympatric*** species, but ecological or phenological[2] barriers mean that wild *Vitis* species remain genetically distinct. Interfertility has long been used by breeders to transfer genes from wild *Vitis* species to *V. vinifera*. Crosses between *Vitis* and *Muscadinia*, on the other hand, are much

2 Climatic impacts on seasonal plant development.

more difficult; nevertheless, they have made it possible to transfer genes to certain grape varieties for resistance to powdery and downy mildews.

Vitis species as a whole are highly heterozygous; in other words, at each ***locus*** of the genome, the sequences of the two chromosomes of the same pair are most often different. The average heterozygosity rate for *V. vinifera*, estimated using ***microsatellite*** markers, is around 75%. Vine was the fourth plant species to have its genome sequenced, first by a Franco-Italian consortium (Jaillon et al. 2007), then by an Italian team who sequenced the Pinot noir grape variety and estimated the differences between the two ***haplotypes*** of an individual at around 2 million *SNPs* for a genome of around 480 million base pairs; that is to say, an average of four SNPs per 1,000 base pairs.

Wild *Vitis* species show a wide variety of forms and ecological adaptability, as they originate in the temperate zones of the northern hemisphere; whereas, the vast majority of species in other genera of the *Vitaceae* family are tropical (Figure 11.2).

11.2.1.3. *The diversity of Vitis vinifera: Wild grapevines, grape varieties, clones and natural mutants*

"Lambrusques" (*Vitis sylvestris*), the wild vine group, are listed as a protected species in France (decree of January 20, 1982[3] establishing the list of protected plant species throughout France). However, few specimens remain in France. Surveys carried out in various locations show the presence of this subspecies in a number of French regions, notably along the Pyrenees mountain range, in Corsica, Languedoc, Charente and Franche-Comté, and in various nature reserves and forests managed by the "Office National des Forêts (National Forestry Service)". Both males and females can be seen, usually in small populations. Males are often covered with flowers at flowering time. Females, on the other hand, show little diversity in their bunches, which are small and loose, and their berries are always small and black, acidic and not very sweet (Figure 11.3(a)).

In contrast to *V. sylvestris*, *V. vinifera* grapes differ widely in bunch and berry size and shape (Figure 11.3(b)), berry skin color and other significant agronomic traits. There is a major difference between table grape varieties (for fresh fruit consumption), with large, often loose bunches and large berries, and wine grape varieties (for wine-making), with smaller, more compact bunches and smaller berries. This diversity allows us to produce a wide variety of products. The organoleptic qualities of wine are not only linked to the identity of the grape variety (or varieties) used to make it, but also to the environmental conditions in which it is

---

3 Available at: https://www.legifrance.gouv.fr/loda/id/JORFTEXT000000865328/.

produced (climate, terrain, etc.), to the production and winemaking methods, and also to the vine's interaction with its ***microbiota*** (in this case, the microorganisms present on or in the plant and around its root system).

Figure 11.2. *Morphological diversity of some Vitis species, in collections or in their natural environment (source: INRAe illustrations). For a color version of this figure, see www.iste.co.uk/pelletier/genetics.zip*

a) b) c)

Figure 11.3. *Example of diversity in berry shape and color in Vitis sylvestris (a), Vitis vinifera (b) and a natural mutant of V. vinifera (c) (source: INRAe illustrations). For a color version of this figure, see www.iste.co.uk/pelletier/genetics.zip*

Another important level of diversity for the wine industry is the variability that exists within a single grape variety, known as intravarietal or clonal variability. Even though repeated cycles of vegetative propagation aim to maintain the same overall characteristics (morphology, product typicity) that define a grape variety's identity, they can still give rise to clonal variation. Clonal variation can relate to a wide range of traits: growth habit, phenology, berry composition in terms of sugars or acids, etc. The vast majority of plant material marketed worldwide is designated as clones of a given grape variety (such as clone 115 of "Pinot noir", clone 214 of "Cabernet franc"). For some grape varieties, there is a very large number of approved clones in France (31 clones of "Cabernet franc", 36 clones of "Chasselas blanc", 38 clones of "Gamay noir", 48 clones of "Pinot noir"), but many grape varieties in the French catalog only have one to five clones[4]. While there is a large number of clones for a grape variety, some clones are used more than others: for example, clones 214, 327 and 623 for "Cabernet franc" and clones 115, 777 and 386 for "Pinot noir" are the most widely used. Grapevines are also relatively sensitive to the appearance of natural mutations, which can affect different organs, sometimes only in part (Figure 11.3(c)). When the mutations involve a quality trait with a strong impact on wine type and quality, the new forms become "essentially derived varieties" of the original variety. This is the case with the different color forms of Pinot ("Pinot noir", "Pinot gris", "Pinot blanc"), or the classic or aromatic form of Savagnin ("Savagnin blanc" and "Gewürztraminer"), which are considered different varieties by growers and authorities, even though, genetically speaking, they are mutants of the same initial grape variety.

Lastly, a further level of diversity is linked to the chimeric state of certain canopy organs (leaves, shoots, fruit). The "Meunier" grape, grown in Champagne, derives from "Pinot noir" by mutating the upper layer of the leaf, which is covered in hairs. Similarly, the gray color of the "Pinot gris" grape is due to the superimposition of two different cell layers: a colored upper cell layer on a white lower layer.

11.3. Origin and domestication of *Vitis vinifera L.*

11.3.1. *Phylogenetics and biogeography of the genus Vitis*

We have pointed out that the various species of the genus *Vitis* are native to Asia, Eurasia and North America. The proximity of *V. vinifera* to Asian *Vitis* species, and therefore the Asian origin of *V. vinifera*, seems to be confirmed by all current research.

4 Available at: https://plantgrape.plantnet-project.org.

However, where is the center of origin of the genus *Vitis*? The distant origin of species in the genus *Vitis* has been the subject of two main hypotheses, supported by molecular biology work. The first hypothesis, supported by ***polymorphism*** analyses of two chloroplast DNA loci and 24 nuclear microsatellite loci among 30 species, points to an Asian origin of the genus *Vitis*, which then dispersed to Europe and America (Péros et al. 2011). The second hypothesis, supported by analyses of polymorphisms in 27 genes scattered across the genomes of 309 ***accessions*** among 48 species and varieties, suggests an American origin of the most recent common ancestor of species in the subgenus *Vitis* (Wan et al. 2013). Differentiation of species in the genus *Vitis* is thought to have begun around –28 million years ago (Mya), with currently existing species formed between –12 and –1.3 Mya.

11.3.2. *Grapevine domestication centers*

When and how did cultivated forms of the *V. vinifera* species emerge? Grapevines were domesticated from their wild ancestors, *V. sylvestris*, of which only a few remain today, as a result of anthropic pressure (deforestation, development of waterways, etc.), but probably also due to the arrival of phylloxera and major pathogens, such as powdery and downy mildews in the 19th century. We can assume that our ancestors first harvested *V. sylvestris* berries in the wild as a supplementary food, and then went on to make wine after inventing pottery. To ensure the supply of grapes, it was then necessary to start cultivating grapevines. We do not know when the first wines appeared, but archaeological remains enable us to estimate the date of wine's appearance on a sufficient scale to be able to truly speak of viticulture. As with many species, the domestication of wine was accompanied by various modifications, commonly referred to as the ***domestication syndrome***. Numerous genetic studies have confirmed the historical stages of grapevine domestication, identifying two primary centers and various secondary domestication centers.

11.3.2.1. *The oldest archaeological evidence of wine production 8,000 years ago*

Did the domestication of the vine and the appearance of cultivated vines precede the production of the first wines, or did the first wines precede the domestication of the vine? Wine production was certainly boosted by the emergence of pottery in the 7th millennium before the present era (BP), enabling food and beverages to be processed and stored. The oldest amphorae discovered, in which the presence of specific wine compounds such as tartaric acid has been confirmed, date back to the Neolithic period (between 5,000 and 6,000 BCE) and originate from sites in Georgia in the South Caucasus region. Given the morphological and biochemical differences between wild and cultivated vines, notably in berry composition (see below), we can assume that the production of wine in large quantities, and with sufficient alcoholic strength to enable the beverage to be preserved properly, required an evolution toward today's cultivated vines. Is the South Caucasian region (Transcaucasia) the area of primary domestication, as these works seem to attest, or would other regions, such as the Fertile Crescent, where many plant species were domesticated, be more likely?

11.3.2.2. *Molecular data supporting the existence of two primary domestication centers and secondary hybridizations*

Numerous molecular studies have been carried out on *V. sylvestris* accessions or by comparing *V. sylvestris* accessions with *V. vinifera* varieties. Numerous hypotheses have been put forward, identifying one or more centers of domestication, in the Mediterranean, the Caucasus or Central Asia. Hybridizations of *V. vinifera* with *V. sylvestris* have also been suggested, identifying secondary centers of domestication.

Very recent and exhaustive work using sequencing data has focused on 1,604 *V. vinifera* grape varieties and 844 *V. sylvestris* accessions from different winegrowing regions around the world (Dong et al. 2023). They confirmed the clear distinction between *V. vinifera* and *V. sylvestris*, and identified four groups of *V. sylvestris* from distinct geographical regions, separated into two ecotypes: the Eastern ecotype, made up of accessions from the Middle East and the Caucasus, and the Western ecotype, made up of accessions from Central Europe and the Iberian Peninsula. Separation between the two ecotypes is thought to have taken place between 200,000 and 400,000 years ago, with the separation of groups within each ecotype dating back 56,000 years for the Eastern ecotype and 2,500 years for the Western ecotype. Based on these groups of *V. sylvestris* diversity, the data generated support the presence of two centers of domestication: one in the Caucasus and the other in the Middle East. Secondary hybridizations between western *V. sylvestris* ecotypes and grape varieties would then have led to the diversification of European grape varieties.

11.4. The main traits that evolved during grapevine domestication

11.4.1. *Domestication syndrome*

The transition from wild sylvestris to today's cultivated grapevine did not occur in a single generation, although the number of generations was undoubtedly fairly limited. Domestication by human selection, whether intentional or not, has led to profound changes in many of the vine's characteristics (This et al. 2006), affecting all its organs as well as its adaptation to the environment. The characteristics most affected are summarized in Table 11.1.

The first concerns the sexual differentiation of flowers, with the transition from dioecia, where male and female flowers are borne by different plants, to hermaphroditism, where the male and female organs present in the same flower ensure better pollination than anemophily between male and female plants and, consequently, more stable production.

The second concerns bunches and berries. The bunches borne by wild female individuals are small and produce few small berries, which are black, often acidic and, above all, not very sweet. Domestication, then selection by humans, has led to great variability in bunch and berry shape, color and composition.

11.4.2. *The presence of male and female organs, a distinctive trait of the wild and cultivated compartments*

In natural conditions, it is possible to find different types of grapevine, both *Vitis vinifera* species and interspecific hybrids, mostly rootstock regrowth in abandoned vineyards or on the edges of vineyards. There are several types of wild *Vitis vinifera* vines that are considered to be "lambrusques": "post-cultural lambrusques", plants whose cultivation has been abandoned, but which have persisted in a wild state; "sub-spontaneous lambrusques", grown in uncultivated soil from cultivated pips; "spontaneous lambrusques", natural elements of the flora which can be divided into three categories; "autochthonous lambrusques", original ancestors of *Vitis vinifera* that have never been cultivated; "colonial lambrusques", resulting from the "sub-spontaneous lambrusques" return to the wild; "mixed lambrusques", resulting from hybridization between autochthonous "lambrusques", and one of the other categories of "lambrusques".

In the wild, it is often difficult to differentiate between these forms. The only "lambrusques" that can be more or less clearly distinguished are post-cultivated "lambrusques", which may retain specific traits of the originally cultivated varieties, particularly in terms of leaf morphology. One of the criteria for differentiating spontaneous "lambrusques" is the gender of the plant (dioecia).

	Trait	*V. vinifera ssp sylvestris*	*V. vinifera ssp sativa*
Plant gender	Presence of male or female organs in the flower.	Separate genders, male and female flowers present on different individuals (dioecia).	Genders grouped together, hermaphroditic flowers bearing both male and female organs. Some rare female varieties.
Fruits	Bunch size and shape.	Bunches are small and loose, with few berries (Figure 11.3(a)).	Bunches are larger and contain a greater number of berries. However, bunches vary widely in size and compactness, depending on their use: table grapes (for fresh consumption) have larger bunches, while wine varieties (for wine production) have smaller, more compact bunches.
	Berry size, shape and color.	Berries are small (weight < 0.5 g), rounded and mostly black.	The berries are larger (from 1–10 g) and come in a wide variety of shapes and colors (Figure 11.3(b)).
	Berry composition and flavor.	Berries have a simple flavor, not very sweet, very acidic.	Sweeter berries, but still highly acidic (sugar/acid balance important for quality); some varieties have Muscat-flavor berries. Variability in the quantity of anthocyanins and tannins.

Table 11.1. *The main traits impacted by domestication and related to production*

11.4.3. *Genetic determinism of flower gender in grapevines*

The vast majority of *Vitis* species are dioecious. The flowers of male plants only have stamens, while the flowers of female plants have a well-developed pistil and so-called reflex stamens, which are non-functional and carry no viable pollen.

Numerous studies have investigated the genetic determinism of gender in grapevine. For example, in *V. sylvestris*, a region of some 150 kb involved in this determinism has been identified on chromosome 2 (Picq et al. 2014). For this region, male individuals would be XY and females would be XX. More recent work has identified a number of genes involved in either male sterility (*VvINP1* (Massonnet et al. 2020); *VviAPRT3* (Badouin et al. 2020)) reflex stamen morphology in female plants (*VviPLATZ* (Iocco-Corena et al. 2021)), or female sterility in male plants (*VviYABBY3* (Massonnet et al. 2020)). All these studies agree that hermaphroditism arose through the recombination between the two regions X and Y, eliminating the male and female sterility genes, an event that could have taken place at least twice independently (Zou et al. 2021).

11.4.4. *Berry color in grapevines: the importance of anthocyanins*

The color of grape berries, and subsequently wine, is due to the presence of color compounds known as anthocyanins. In most grape varieties, anthocyanins are only present in the skin. In rare grape varieties, known as "Teinturiers", anthocyanins are also present in the pulp, in which case, they are either synthetized directly in the pulp ("Alicante Henri Bouschet" grape variety), or are diffused from the skin when a very large quantity of anthocyanin is produced there.

11.4.4.1. *Presence or absence of anthocyanins*

The "white berry" trait is due to inhibition of expression of the *VvMybA1* gene, a ***transcription***[5] factor that regulates expression of the *VvUFGT* gene, a key gene in the anthocyanin biosynthesis pathway (Kobayashi et al. 2004). This inhibition is due to the insertion of a retro-transposon[6], GRET1, into the gene ***promoter***. When *VvMybA1* is no longer expressed, due to the insertion of GRET1, anthocyanins are no longer synthesized. As the mutation is recessive, white-berry varieties are homozygous for the insertion, and the black-berry varieties are homozygous or heterozygous for the wild gene, although the vast majority of black varieties are heterozygous. The gray color of certain grape varieties, such as "Pinot gris", is due to a chimeric state of the skin: the first cell layer of the skin is black, while the second cell layer is white.

11.4.4.2. *Variation in color intensity*

The color intensity of the grape berry skin is highly variable, from very pale berry varieties such as "Gewürztraminer", through pink, red and gray berry varieties, to dark berry varieties such as the "Syrah". The quantity of anthocyanins present in

5 Type of protein required to initiate or regulate gene transcription.

6 A ***transposon*** whose propagation involves an RNA intermediary retro-transcribed into DNA molecules, enabling it to multiply in the genome.

the skin explains this variation. A first region of the genome involved in the control of this trait on chromosome 2, at the level of a group of three *Myb* genes[7], *VvMybA1*, *VvMybA2*, *VvMybA3* (Fournier-Level et al. 2009), and a second region at the level of the *VvUFGT* gene promoter (Huang et al. 2013) have been identified. More recently, nine additional regions have also been identified on chromosomes 2, 4, 6, 11 and 17.

GRET1 can also excise from its position upstream of the *VvMybA1* gene. If the retro-transposon is fully excised, there is a reversion to black berries. However, it is also possible for the GRET1 excision to not be completed, and the presence of small fragments of GRET1 upstream of *VvMybA1*, or an insertion in an intron, leads to certain colors: pink or red (Shimazaki et al. 2011). The *VvMybA* locus appears to be relatively unstable, as excision of a large part of the locus can also be observed in white berry grape varieties (Walker et al. 2002), or a recombination between *VvMybA1* and *VvMybA3* accompanied by a 69-kb deletion that restores *VvMybA1* transcripts, and hence color (Röckel et al. 2022). This locus instability is the major cause of somatic variation for this trait. Epigenetic mechanisms involved in color control have also been demonstrated: the methylation[8] level of the *VvMybA1* promoter or of the retro-transposon GRET1 (Azuma and Kobayashi 2022) can thus regulate skin coloration. Lastly, to be complete, it should also be noted that berry color is influenced by the environment, including temperature and sunlight (Azuma et al. 2012) or other characteristics, such as rootstock.

11.4.4.3. *Color variation*

The color of the skin, therefore ultimately the color of the wine, depends on the anthocyanins present in the skin. There are five main anthocyanins, in their mono-glucoside and acyl-mono-glucoside forms: delphinidin, cyanidin, peonidin, petunidin and malvidin. These five anthocyanins differ from one another in the number and position of hydroxyl and methoxyl groups: delphinidin and cyanidin are not methylated, peonidin and petunidin are mono-methylated and malvidin is di-methylated. Increasing the number of methyl groups on the molecule changes its color from red to blue. Two regions of the genome involved in the control of this feature, one corresponding to the *Myb* gene cluster on chromosome 2, the other corresponding to an O-methyl-transferase gene cluster, have been identified

---

7 A family of transcription factors that specifically recognize a (C/T)AAC(G/T)G DNA sequence, and are defined by the presence of a domain of approximately 50 amino acids containing three tryptophans.

8 Biological process by which methyl groups (CH_3) are added to cytosines in the DNA molecule, potentially modifying its activity.

(Fournier-Level et al. 2011; Azuma et al. 2015). The *Myb* locus could have a regulatory effect on the expression level of genes encoding flavonoid 3′,5′-hydroxylase and flavonoid 3′-hydroxylases, two enzymes involved in the anthocyanin biosynthetic pathway.

11.5. From domestication to the present day

From the area of primary domestication, cultivated grapevines followed the main civilizations and sea routes up to our era, moving globally toward western Europe. Numerous illustrations, dating from the period of Egyptian, Greek, Cretan and Roman civilizations, show scenes celebrating vines. Dionysus was the Greek god of wine and the grape harvest, as well as Bacchus, his Roman equivalent. At the same time, table and wine grapes were differentiated. On the other hand, it is very difficult to find ancient writings describing the plant material used, particularly grape varieties.

The berries of most *Vitis* species are black, as are those of *V. sylvestris*. White and intermediate colors (pink, red) have only appeared in cultivated *V. vinifera*. Molecular diversity data on the *VvMybA1-2-3* gene cluster (see above) are consistent with a more ancient origin of black grape varieties; white grape varieties have a more recent origin and rapid spread that is consistent with the spread of winemaking after the decline of the Roman Empire (Fournier-Level et al. 2010).

11.5.1. *Ancient grape varieties*

The first grape varieties were described by famous Roman authors such as Cato the Elder, Varron, Virgil, Pliny the Elder and Columella. Although grape variety names appear in these works, it is not possible to link them to today's grape varieties. The first names of grape varieties that could correspond to our present-day varieties appear from the 13th century onward; the following are examples of this: in his treatise on rural economics *Opus ruralium commodorum*, written in 1303, Pierre de Crescence (Pietro Crescenzi) cites certain grape varieties, such as "Muscat". Estienne and Liébault, in 1572, in their book *L'Agriculture et Maison rustique*, also cite "Pinot", "Gouais" and "Meslier" grape varieties. Rabelais, in *Gargantua et Pantagruel*, cites the "Chenin" variety "with large chenin berries". Lastly, Olivier de Serres, in *Le Théâtre d'agriculture et mesnage des champs*, cites "Malvoisie", "Bourboulenc" and "Chatus".

Even if we cannot be certain that these names refer to the same grape varieties as today, they nevertheless attest to the presence of well-defined grape varieties, probably as early as the Roman period. How these appeared and have been preserved to the present day is a question to which a great deal of work has been devoted. In a recent study, Ramos-Madrigal et al. (2019) identified a first-century grape seed whose genetic profile, compared with a reference sequence database, corresponds to a direct descendant of the "Savagnin blanc" grape variety. Similarly, samples dating back to the 11th century correspond completely to this same "Savagnin blanc", demonstrating 1,000 years of vegetative propagation of this grape variety.

11.5.2. *Relationships between grape varieties*

The use of microsatellite markers has shown that many present day grape varieties are the result of natural crosses: "Cabernet Sauvignon", as its name suggests (although it was named long before this demonstration), is the result of a cross between "Cabernet franc" and "Sauvignon". Similarly, many famous grape varieties, such as "Chardonnay", "Gamay" and "Aligoté", are the result of crosses between "Pinot noir" and "Gouais blanc", a variety that has all but disappeared from French vineyards (Bowers et al. 2002). Numerous studies have now described the relationships between grape varieties. Two important conclusions can be drawn from them:

– today, it is not possible to go back beyond a limited number of generations: a study of the Domaine de Vassal[9] collection, comprising almost 2,300 grape varieties from some forty countries, has revealed similarities in no more than seven interconnected generations (Lacombe et al. 2013);

– varietal genealogies reveal very frequent overlapping generations (Figure 11.4).

Generation overlap is made possible by the maintenance of a given grape variety by vegetative propagation over a very long period (several centuries for some, such as "Pinot noir", which may already have been present in Roman times, or "Savagnin", mentioned above). It is, however, more difficult to understand the small number of interconnected generations identified to-date. The loss of diversity that accompanied the phylloxera crisis (see below) undoubtedly resulted in the disappearance of many grape varieties, particularly those that were little used.

9 Available at: https://vassal.montpellier.hub.inrae.fr/.

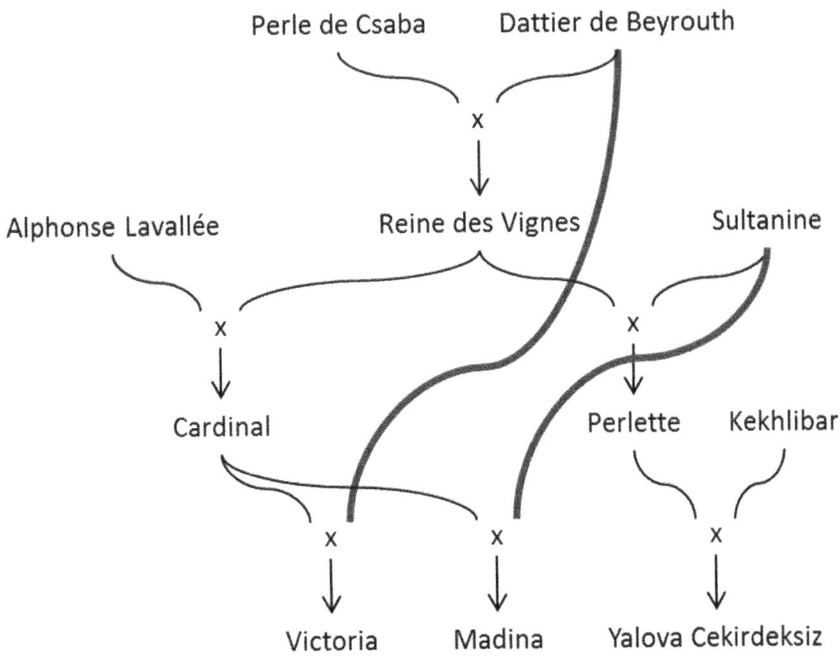

Figure 11.4. *Example of generation overlap: Dattier de Beyrouth, which is a parent of Reine des vignes and of Victoria, is itself descended from Reine des vignes (source: Lacombe et al. 2013). For a color version of this figure, see www.iste.co.uk/pelletier/genetics.zip*

11.5.3. *Table grapes, wine grapes*

Early authors classified cultivated grapevines into three eco-geographical groups: *Orientalis*, *Pontica* and *Occidentalis*, the latter being considered the ancestral group. A study of the grape varieties in Vassal's collection also identified three gene pools: a group of wine grape varieties originating in the Balkans and Eastern Europe as far as the Caucasus, a group of wine grape varieties from the West and a group of table grape varieties from Eastern Europe.

The recent study by Dong et al. (2023) finds no clear association between the geographical origin of grape varieties and their genetic structure, probably due to the exchange of varieties between regions and the subsequent cross-breeding that has taken place over the course of history. It identifies six groups corresponding to different genetic origins, pre-dating the genetic pools identified in the Vassal collection: two ancestral groups (Middle Eastern table grape varieties and Caucasian wine grape varieties) and four more recent groups (Muscat grape varieties, wine grape varieties from the Balkans, the Iberian Peninsula and Western Europe). The two ancestral groups, which probably appeared simultaneously, would represent the two centers of domestication: the Middle East and the Caucasus. The authors estimate the period of separation of the populations to be around 11,000 years ago. Ultimately, the data show that the Caucasus wine grape group was rather local and had little influence on the diversification of grape varieties. On the other hand, groups from the Middle East domestication center dispersed, particularly in Africa and Eurasia, strongly influencing diversification. These table grape varieties would, on the one hand, have followed a route south of the Mediterranean, and, on the other hand, eastwards to Central Asia, along land-based trade routes. More recently, the expansion of wine grape varieties has been directly associated with the spread of Christianity, which used wine in religious services. This is how the grapevines left the shores of the Mediterranean and moved northwards, following the progress of the Church. Similarly, following the colonization of the New World, grapevines were imported to America by the Jesuits in the 17th century. The first grapevines imported as cuttings were crossed on the spot, giving rise to the "Criolla" grape varieties. Then, other waves of settlers (French, German, etc.) imported shoots, replicating the Old-World grape varieties identically (e.g. "Cot" = "Malbec", "Carménère", etc.).

11.5.4. *The phylloxera crisis and its consequences*

A fundamental event in the development of winemaking and the evolution of grape varieties was the phylloxera crisis (Pouget 1990). Originating in America and detected in France as early as 1863, phylloxera (*Daktulosphaira vitifoliae*), an insect related to aphids, was responsible for the near total disappearance of European vineyards. Its radicicolous form attacks the roots of the vine: they sting the roots, resulting in the formation of tuberosities. Necrosis then develops, damaging the roots and allowing various microorganisms to enter, which are responsible for rot, which leads to vine death within three years. Various control methods were tested, but grafting onto phylloxera-resistant rootstock was ultimately what was successfully developed. This is the greatest historical example of biological control of a plant pest.

Numerous breeders in the late 19th and early 20th centuries really initiated the period of directed grapevine selection. The selection of rootstocks, resistant to phylloxera and presenting other characteristics of adaptation to different types of soil (notably limestone), began with the selection of a few accessions of wild species, such as "Rupestris du Lot" or "Riparia Gloire de Montpellier". This was soon followed by interspecific crosses between various species, which were mainly American. A new profession emerged to multiply plant material during the "post-phylloxera reconstitution": grapevine nurserymen. The period of rootstock creation was followed by a period of variety creation known as "hybrid direct producers" ("*hybrides producteurs directs*" (HPD)). These are varieties resistant to phylloxera, powdery mildew and downy mildew (two other vine pests that followed the same path as phylloxera), and produce grapes, planted on their own roots. The first hybrid grape varieties were highly resistant to powdery mildew and downy mildew, but of mediocre quality. Subsequent generations improved wine quality to the detriment of disease resistance. In the 1960s, half of French vineyards were planted with hybrid grapes, but today they have all but disappeared from the vineyard.

The creation of *Vitis vinifera* varieties also continued in the late 19th and 20th centuries, with table grape varieties such as "Italia" and wine grape varieties such as "Muscat Ottonel", "Alicante H. Bouschet" and "Marselan". For table grape varieties, an increasingly sought-after seedless trait has led to numerous varieties descended from "Sultanina" (or Thompson seedless in the United States), a natural seedless mutant, such as "Centennial Seedless" or "Flame Seedless".

11.6. The grapevine of tomorrow

As we have seen, the history of the vine has been intimately linked to human history, with humans strongly influencing its evolution. The same is very likely to be true in the future. Winegrowers are increasingly demanding varieties and cultivation methods that reduce phytopharmaceutical inputs, as called for by public authorities, consumers and local residents. Northern wine-producing countries (Germany, Switzerland and Hungary) initiated the trend, and today all wine-producing countries are involved in the creation of grape varieties that are resistant to the main vine diseases, notably powdery mildew and downy mildew. Furthermore, future climate trends will have a major impact on production levels and product quality, reducing the sustainability of vineyards. Some vineyards have begun experimenting with grape varieties from more southerly regions, for example, Greek or Spanish varieties in the vineyards of Bordeaux or Occitanie. Varietal creation has also taken up this challenge and may soon produce grape varieties that are better adapted to more severe drought or higher temperatures. It is likely that the winegrowing landscape of the future will be planted with a new generation of varieties and will invest in new territories.

11.7. References

Azuma, A. and Kobayashi, S. (2022). Demethylation of the 3' LTR region of retrotransposon in VvMYBA1^{BEN} allele enhances anthocyanin biosynthesis in berry skin and flesh in 'Brazil' grape. *Plant Science*, 322, 111341. doi: 10.1016/j.plantsci.2022.111341.

Azuma, A., Yakushiji, H., Koshita, Y., Kobayashi, S. (2012). Flavonoid biosynthesis-related genes in grape skin are differentially regulated by temperature and light conditions. *Planta*, 236, 1067–1080. doi: 10.1007/s00425-012-1650-x.

Azuma, A., Ban, Y., Sato, A., Kono, A., Shiraishi, M., Yakushiji, H., Kobayashi, S. (2015). MYB diplotypes at the color locus affect the ratios of tri/di-hydroxylated and methylated/non-methylated anthocyanins in grape berry skin. *Tree Genetics & Genomes*, 11, 31. doi: 10.1007/s11295-015-0855-0.

Badouin, H., Velt, A., Gindraud, F., Flutre, T., Dumas, V., Vautrin, S., Marande, W., Corbi, J., Sallet, E., Ganofsky, J. et al. (2020). The wild grape genome sequence provides insights into the transition from dioecy to hermaphroditism during grape domestication. *Genome Biology*, 21, 223. doi: 10.1186/s13059-020-02131-y.

Bowers, J., Boursiquot, J.M., This, P., Chu, K., Johansson, H., Meredith, C. (2002). Historical genetics: The parentage of chardonnay, gamay, and other wine grapes of Northeastern France. *Science*, 285, 1562–1565. doi: 10.1126/science.285.5433.1562.

Dong, Y., Duan, S., Xia, Q., Liang, Z., Dong, X., Margaryan, K., Musayev, M., Goryslavets, S., Zdunić, G., Bert, P. et al. (2023). Dual domestications and origin of traits in grapevine evolution. *Science*, 379, 892–901. doi: 10.1126/science.add8655.

Fournier-Level, A., Le Cunff, L., Gomez, C., Doligez, A., Ageorges, A., Roux, C., Bertrand, Y., Souquet, J.M., Cheynier, V., This, P. (2009). Quantitative genetic bases of anthocyanin variation in grape (*Vitis vinifera* L. ssp sativa) berry: A QTL to QTN integrated study. *Genetics*, 183, 1127–1139.

Fournier-Level, A., Lacombe, T., Le Cunff, L., Boursiquot, J.M., This, P. (2010). Evolution of the VvMybA gene family, the major determinant of berry colour in cultivated grape-vine (*Vitis vinifera* L.). *Heredity*, 104(4), 351–362.

Fournier-Level, A., Hugueney, P., Verries, C., This, P., Ageorges, A. (2011). Genetic mechanisms underlying the methylation level of anthocyanins in grape (*Vitis vinifera* L.). *BMC Plant Biology*, 11, 179.

Huang, Y.F., Bertrand, Y., Guiraud, J.L., Vialet, S., Launay, A., Cheynier, V., Terrier, N., This, P. (2013). Expression QTL mapping in grapevine—Revisiting the genetic determinism of grape skin colour. *Plant Sci.*, 207, 18–24.

Iocco-Corena, P., Chaïb, J., Torregrosa, L., Mackenzie, D., Thomas, M.R., Smith, H.M. (2021). VviPLATZ1 is a major factor that controls female flower morphology determination in grapevine. *Nature Communications*, 12, 6995. doi: 10.1038/s41467-021-27259-8.

Jaillon, O., Aury, J.M., Noel, B., Policriti, A., Clepet, C., Casagrande, A., Choisne, N., Aubourg, S., Vitulo, N., Jubin, C. et al. (2007). The grapevine genome sequence suggests ancestral hexaploidization in major angiosperm phyla. *Nature*, 449, 463–467.

Kobayashi, S., Goto-Yamamoto, N., Hirochika, H. (2004). Retrotransposon induced mutations in grape skin color. *Science*, 304, 982.

Lacombe, T., Boursiquot, J.M., Laucou, V., Di Vecchi-Staraz, M., Péros, JP., This, P. (2013). Large-scale parentage analysis in an extended set of grapevine cultivars. *Theoretical and Applied Genetics*, 126, 401–414.

Massonnet, M., Cochetel, N., Minio, A., Vondras, A.M., Lin, J., Muyle, A., Garcia, J.F., Zhou, Y., Delledonne, M., Riaz, S. et al. (2020). The genetic basis of sex determination in grapes. *Nature Communications*, 11, 2902. doi: 10.1038/s41467-020-16700-z.

Péros, J.P., Berger, G., Portemont, A., Boursiquot, J.M., Lacombe, T. (2011). Genetic variation and biogeography of the disjunct Vitis subg. Vitis (Vitaceae). *Journal of Biogeography*, 38, 471–486.

Picq, S., Santoni, S., Lacombe, T., Latreille, M., Weber, A., Ardisson, M., Ivorra, S., Maghradze, D., Arroyo-Garcia, R., Chatelet, P. et al. (2014). A small XY chromosomal region explains sex determination in wild dioecious V. vinifera and the reversal to hermaphroditism in domesticated grapevines. *BMC Plant Biology*, 14, 229. doi: 10.1186/s12870-014-0229-z.

Pouget, R. (1990). *Histoire de la lutte contre le phylloxera de la vigne en France*. INRA, Paris.

Ramos-Madrigal, J., Runge, A.K.W., Bouby, L., Lacombe, T., Samaniego Castruita, J.A., Adam-Blondon, A.F., Figueiral, I., Hallavant, C., Martínez-Zapater, J.M., Schaal, C. et al. (2019). Palaeogenomic insights into the origins of French grapevine diversity. *Nat. Plants*, 5, 595–603. doi: 10.1038/s41477-019-0437-5.

Röckel, F., Moock, C., Schwander, F., Maul, E., Töpfer, R., Hausmann, L. (2022). A 69 kbp deletion at the berry color locus is responsible for berry color recovery in *Vitis vinifera* L. cultivar ‘Riesling Rot’. *Int. J. Mol. Sci.*, 23, 3708. doi: 10.3390/ijms23073708.

Shimazaki, M., Fujita, K., Kobayashi, H., Suzuki, S. (2011). Pink-colored grape berry is the result of short insertion in intron of color regulatory gene. *PLoS ONE*, 6(6), e21308. doi: 10.1371/journal.pone.0021308.

This, P., Lacombe, T., Thomas, M.R. (2006). Historical origins and genetic diversity of wine grapes. *Trends in Genetics*, 22(9), 511–519.

Walker, A.R., Lee, E., Robinson, S.P. (2006). Two new grape cultivars, bud sports of Cabernet Sauvignon bearing pale-coloured berries, are the result of deletion of two regulatory genes of the berry colour locus. *Plant Molecular Biology*, 62, 623–635. doi: 10.1007/s11103-006-9043-9.

Wan, Y., Schwaninger, H.R., Baldo, A.M., Labate, J.A., Zhong, G.Y., Simon, C.J. (2013). A phylogenetic analysis of the grape genus (*Vitis* L.) reveals broad reticulation and concurrent diversification during neogene and quaternary climate change. *BMC Evol. Biol.*, 13, 141. doi: 10.1186/1471-2148-13-141.

Zou, C. Massonnet, M., Minio, A., Patel, S., Llaca, V., Karn, A., Gouker, F., Cadle-Davidson, L., Reisch, B., Fennell, A. et al. (2021). Multiple independent recombinations led to hermaphroditism in grapevine. *Proc. Natl. Acad. Sci. U.S.A.*, 118. doi: 10.1073/pnas.2023548118.

12

Tomato Domestication and Breeding: A Major Contribution from Wild Species

Mathilde CAUSSE
GAFL, INRAE, Avignon, France

12.1. Introduction

Thanks to its characteristic shape, color and taste, the tomato is one of the world's most popular vegetables. It is appreciated in every country for its relative ease of cultivation, high yields and diverse uses, both fresh and processed. Tomato production and yields have risen dramatically since it became an international success in the 20th century. In 2019, global tomato production stood at around 180 million tons (MT), compared with 30 MT in the early 1960s and 109 MT in 2000 (source: FAOstat). China and the United States are the world's leading tomato producers, accounting for 33% and 8% respectively. On the European continent, France ranks 21st (around 600,000 tons), far behind Italy, Spain and Greece. After a brief presentation of the plant, we will describe the evolution of knowledge concerning its origin and domestication, and show the significance of South American resources for modern tomato breeding. New knowledge acquired in recent years, thanks to advances in genomics, will also be detailed.

Genetics of Domestications,
coordinated by Georges PELLETIER.

12.1.1. *Botanical and agronomic description*

The tomato is a herbaceous plant grown for its fruit, which is a low-calorie, water-rich diet food. It contains many minerals and vitamins, the most important of which is vitamin C, as well as pigments from the carotenoid family. Lycopene, which is responsible for the fruit's red color, plays an antioxidant role in the human diet, as well as in the prevention of certain cardiovascular diseases and cancers.

The tomato is a perennial plant, but is grown as an annual plant. It is an indeterminate-growth[1] plant, but there are certain determinate-growth[2] varieties that have enabled the development of mechanized harvesting, notably for industrial tomatoes, while other varieties, which must be staked and harvested by hand, can produce for several months. In Europe, crops for the fresh produce market are mainly grown under cover.

12.1.2. *Etymology and systematics*

The tomato belongs to the Solanaceae family and the genus *Solanum*. This family comprises several thousand recorded species and includes food plants such as potatoes (*Solanum tuberosum*), eggplants (*Solanum melongena*) and chilies (*Capsicum annuum*), as well as tobacco (*Nicotiana tabacum*) and ornamental plants such as jasmine nightshade (*Solanum jasminoides*). Some species in the Solanaceae family are known for their toxicity, such as black nightshade (*Solanum nigrum*) or bittersweet (*Solanum dulcamara*), while others are known for their psychotropic properties, such as mandrake (*Mandragora officinarum*) or belladonna (*Atropa belladonna*).

The tomato has been known by many names. In Mexico City, the Nahuatl-speaking people called the cherry tomato *tomatl*, from which the name "tomato" was derived. In the 17th century, it took on the vernacular name of *Mala aura* or *Mala peruviana*, in Italian *Pomi d'oro*, and in Spanish *Pomi des Moro*. As soon as the tomato was introduced to Europe, botanists included this species in the genus *Solanum*, like the potato. In 1694, Joseph Tournefort was the first to distinguish the cultivated tomato and create a new genus, *Lycopersicum*, to classify

1 The tomato is a liana that grows by producing 5–12 leaves, then a floral bunch, then three leaves and a bunch, continuously: this is called indeterminate growth. Some varieties produce a few floral bunches, then stop growing: this is known as determinate growth. These varieties are particularly used for the production of canned tomatoes, which are grown in fields and harvested mechanically in a single pass.

2 *Ibid.*

this species. The species name *Esculentum* comes from Latin and means "edible". The systematics of the genus *Lycopersicon* has evolved over the years. Botanical criteria were first used to distinguish the different species, and more recently, molecular criteria have been used to confirm and refine a classification that has long been accepted and used by geneticists. Although tomatoes were, until recently, recognized as belonging to the genus *Lycopersicon* by most taxonomists, their classification within the genus *Solanum* is now widely accepted. It is supported by the converging results of several ***phylogenetic*** studies based on morphological and molecular criteria. The cultivated species *Solanum lycopersicum* can be divided into two botanical groups: *esculentum* and *cerasiforme*. These two groups differ in terms of fruit size (between 1.5 and 2.5 cm for *cerasiforme*, corresponding to cherry tomato types) and number of fruit locules (two locules for *cerasiforme* and two or more locules for *esculentum*). The cherry tomato, *S. lycopersicum var. cerasiforme*, was early described as the domesticated ancestor of the cultivated tomato.

12.2. Origin of the cultivated tomato: wild ancestors and centers of domestication

Figure 12.1. *Shape and color diversity of tomato varieties (source: C. Slagmulder, INRAe). For a color version of this figure, see www.iste.co.uk/pelletier/genetics.zip*

Species	Geographical distribution and habitat	Mode of reproduction	Fruit color	Genetic diversity
S. arcanum	Northern Peru, 100–2,500 m, dry environment	SI, optional AL	Green + dark streaks	Medium
S. chmielewskii	Southern Peru – northern Bolivia, 1,500–3,000 m, dry environment	AC	Green + dark streaks	Medium
S. neorickii	Southern Ecuador – southern Peru, 1,500–3,000 m, wetlands	AC, AF	Green + dark streaks	Low
S. pennellii	Peruvian coast, 0–2,000 m, dry environment	SI and SC, AL	Green	High
S. habrochaites	Southwestern Ecuador – southern Peru, 500–3,000 m, rainforest	SI and SC, AL	Green + dark streaks	High
S. chilense	Southern Peru – northern Chile, 0–3,000 m, dry environment	SI, AL	Light green	High
S. huaylasense	Peru, 1,700–3,000 m, rocky slopes	SI, AL	Green + dark streaks	High
S. peruvianum	Central Peru – northern Chile, 0–600 m, coastal zone	SI, AL	Green to white	High
S. corneliomulleri	Southern Peru, 1,000–3,000 m	SI, AL	Green + dark streaks	High
S. cheesmaniae	Galapagos Islands, 0–1,300 m, coastal and volcanic zone	SC, AF	Yellow, orange	Low
S. galapagense	Galapagos Islands, coastal zone	SC, AF	Yellow, orange	Low
S. pimpinellifolium	Southern Ecuador – northern Peru, 0–1,000 m	SC, AF and AL	Red	Medium
S. lycopersicum var. cerasiforme	Probable origin Ecuador – Peru, now everywhere	SC, AF	Red	Weak
S. lycopersicum	Probable origin Ecuador – Peru – Mexico, now everywhere	SC, AF	Red	Very low

Table 12.1. *Main characteristics of species in the Lycopersicon section of the genus Solanum (SI: self-incompatible; AL: allogamous; SC: self-compatible; AF: autogamous)*

The process of domestication implies that a species acquires new characteristics compared to a wild ancestor, which give it an advantage in use by human communities. The ***domestication syndrome*** generally includes a more compact plant habit, earlier flowering, reduction or loss of seed dormancy and dispersal, gigantism and an increase in the diversity of parts consumed. In plants where the fruit is the consumed part, the traits influencing fruit morphology are the main criteria that have evolved with domestication. Tomato domestication led to an increase in the diversity of fruit shape and color, and an increase in fruit size (Bai and Lindhout 2007; Figure 12.1). Humans have also selectively bred the tomato's ability to produce identical plants in subsequent generations by establishing ***autogamy***.

The 12 wild tomato relatives are distributed along the Andean coast from Ecuador to northern Chile, with two species that are endemic to the Galapagos Islands (Table 12.1). Although the wild relatives are now well known in tomatoes, the domestication process was controversial until recently. This process is undoubtedly very old. The tomato was already widely cultivated in Mexico before the Spanish conquest. It was probably also found in Peru. Indeed, both wild and cultivated plants have been found in Peru, and the term *tomato* comes from the island of Hispaniola (present-day Haiti and Santo Domingo). In order to understand the origin of a cultivated plant, several factors need to be considered. These include botany (geographical distribution of related species), genetic diversity, archaeology, history and linguistics.

12.2.1. *Indications provided by botany*

Nikolai Ivanovich Vavilov (1887–1943) was the first to describe the "centers of diversity" of cultivated plants, the geographical areas where cultivated plant species first develop their distinctive traits. He believed that the first stages in a plant's domestication took place within these primary centers. Transportation out of these centers was then due to human migration. The fact that most wild tomato species are endemic to South America, and restricted to a limited Andean zone, favors the hypothesis of domestication in this region. Nikolai Vavilov admits, however, that domestication may have taken place far from the center of primary diversity in plants that were domesticated at a later stage, with a gradual expansion of wild species into areas occupied by human activity. In the case of tomatoes, while there is no question of doubting the center of origin[3] of the genus *Solanum sect.*

3 Region where domestication of a species took place. In the case of plants, it often corresponds to the region of diversification of the wild species.

Lycopersicon, it is not certain that the center of diversity of the cultivated tomato (likely area of domestication) is the same as the center of diversity of wild relatives. The Mexico City region is the presumed center of origin of the cultivated species. Given that the wild *S. pimpinellifolium*, found between Peru and Ecuador, is the closest wild species to the cultivated species, the tomato ***accessions*** introduced into Mexico must nevertheless have undergone early stages of domestication. Indeed, it would be astonishing for little red berries to travel 5,000 km (from the Andean regions to Mexico) without receiving the slightest attention from the humans who encountered or transported them.

12.2.2. *Archaeological and historical evidence*

No archaeological data have been found that can remotely be linked to tomato cultivation. No remains of tomato plants or seeds have been found in archaeological digs in Peru, Ecuador or even Mexico. The only potential representation found in Colombia is a wool-spinning spool decorated with a tomato flower, dating from 900 to 500 BCE. However, the interpretation of this spool is disputed. The absence of tomato representations left by the various pre-Columbian civilizations reflects the fact that this plant must not have been of major interest in their diet, unlike maize and chili for the Aztecs or potatoes for the Incas.

While there are no written records of tomato cultivation in South America, information can be found in writings at the time of the Spanish conquest of Mexico. Bernardino de Sahagún collected the testimonies of Aztecs who had known native society before the Spanish conquest. He transcribed these accounts in *Histoire générale des choses de la Nouvelle-Espagne*. The tomato is described as a common Aztec food. In particular, it was used as a medicinal plant and seemed to be widespread, both in religious offerings and in markets. It was probably used as a condiment with chili pepper in sauces. It was also used to garnish cannibal dishes, if we are to believe the text describing Hernan Cortes' conquest of Mexico. It seems that the first tomatoes, introduced to Europe from Mexico, showed great diversity and had already undergone an advanced stage of domestication, with the appearance of medium-sized fruit with numerous locules and a yellow color, indicating a secondary site of domestication.

12.2.3. *Indications provided by linguistics*

The word "tomato" is said to derive from the Spanish word *tomate* (pronounced *tomaté*), itself derived from the suffix *tomatl*, used in the Nahuatl language (spoken by the Aztecs in Central America and still used in Mexico). The word *xitomatl* seems to refer to the tomato as we know it in Europe, and it is used by Fernando Alvarado Tezozomoc (1598) in *Cronica mexicayotl* to describe the cultivated tomato. Today, in Mexico City, cultivated tomatoes are still called *xitomatl*, *gitomate* or *jitomate*, the latter two names probably deriving from the former. The Mayas (further south) use a different etymology (with the prefix *p'aak*) and have never adopted the Nahuatl names. The fact that such different names are used in different ethnic regions seems to support an early introduction of the tomato to Central America. It is also rather unsatisfactory to conclude that the tomato originated from the names given to the plant by Europeans. Indeed, the terms *Pomi del Peru* and *Mala peruviana* suggest a Peruvian origin, but they were previously used in France to name Datura (*Datura stramonium*). There may also have been confusion over the term *Lycopersicon*, used by Galen in the 2nd century to designate Datura or eggplant, even though tomatoes did not exist in the Old World.

12.3. The origins of cultivated tomatoes: genetic data

12.3.1. *Indications provided by genetic diversity*

Charles Rick (1915–2002), a professor at the University of California, Davis, devoted his career to studying the origin and genetics of the tomato. On the basis of a few ***isoenzyme markers***, he hypothesized that domestication had taken place from wild accessions of *Solanum lycopersicum var. cerasiforme* that had evolved directly from wild species (Rick 1976). He even suggested that the species *Solanum pimpinellifolium* may have come from a parallel branch during the evolution of the tomato. However, analyses of the relationships between morphological traits and ***molecular markers*** show a direct filiation between *S. pimpinellifolium*, *S. l. cerasiforme* and *S. l. esculentum*. Today, *S. l. cerasiforme* is considered a primitive type of cultivated tomato or a transitional form between wild *S. pimpinellifolium* and cultivated *S. lycopersicum*. It would also appear that many cherry tomato accessions are of ***feral*** origin.

The range of *S. l. cerasiforme* extends to all tropical and subtropical regions of the world. This post-Columbian expansion erases the traces of the species' historical diffusion during domestication. The adventitious nature of *S. l. cerasiforme* may explain why this species is so often associated with man. For *S. l. cerasiforme* and *S. l. esculentum*, greater genetic variability was found for accessions from Peru and Ecuador than for accessions from Mexico. In these regions, where ***sympatry*** with

S. pimpinellifolium is common, *S. l. cerasiforme* and *S. l. esculentum* accessions share ***alleles*** that are derived from *S. pimpinellifolium*. These shared alleles reflect recent intercrossing events, rather than potential kinship. The fact that *S. l. cerasiforme* accessions surveyed in the Andean zone are more variable than extra-Andean accessions supports the idea that tomatoes were domesticated from Andean populations of *S. l. cerasiforme*. As the cherry tomato is a morphological intermediate between wild *S. pimpinellifolium* and cultivated *S. l. esculentum*, it can be assumed that domestication took place from *S. pimpinellifolium* to *S. l. cerasiforme*, then to *S. l. esculentum*. Nevertheless, recurrent intercrossing between the wild species and botanical varieties also took place subsequently. Other authors believe that the missing link between the wild species and the cultivated tomato would be a species growing northeast of Ecuador, in a humid environment. They describe a new species: *Lycopersicon humboldtii*, which has only been identified twice and strongly resembles *S. l. cerasiforme*.

12.3.2. *Genomics and its contribution*

Since the advent of molecular markers in the late 1980s, studies have shown the drastic reduction in diversity between wild species that are close to tomato and the cultivated species. The close proximity between *S. lycopersicum* and *S. pimpinellifolium* was confirmed, along with the intermediate status of *S. l. cerasiforme*. However, it was only with the identification of DNA sequence polymorphisms, first of genes and then of the whole genome (TGC 2012), that research was able to make progress on these issues.

The sequencing of the tomato genome was carried out by a consortium of 14 countries and began in 2003, before finally being published in 2012 (TGC 2012). Initially, there was a plan in place to only sequence the gene-rich regions (around 250 Mb), but with the advent of next-generation sequencers, almost the entire genome (760 of the 950 Mb) could be sequenced and ordered. Comparison between the genomes of *S. lycopersicum* and *S. pimpinellifolium* reveals a divergence of around 0.6%, confirming the proximity of the two species. The order, orientation and location of the 35,000 genes were obtained and compared with those of a potato, revealing the close proximity of the two species.

Very quickly, numerous varieties were resequenced and their genomes compared with that of the reference genome (Lin et al. 2014). This work identified regions showing traces of natural or selection-induced breeding. Lin et al. (2014), for example, resequenced 360 accessions, including 53 *S. pimpinellifolium*, and identified over 10 million polymorphisms. Comparison of wild and cultivated accessions identified 186 regions showing a reduction in diversity as a result of domestication and 133 regions impacted by further selection (Figure 12.2).

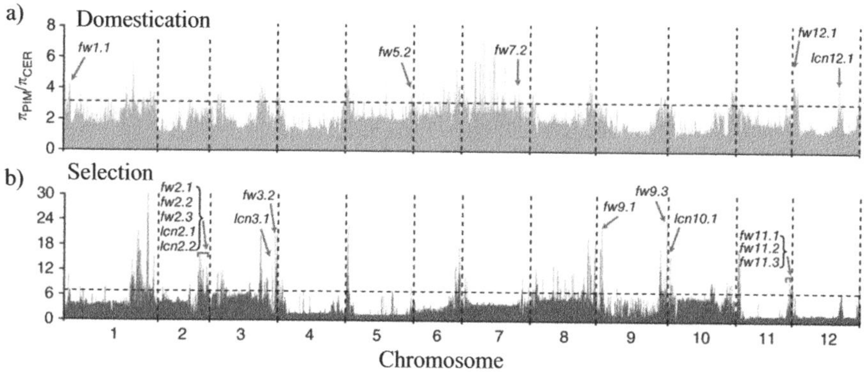

Figure 12.2. *Relationship between the diversity of wild forms and cherry tomatoes (a) and cherry tomatoes and large-fruited varieties (b) along the 12 chromosomes. Peaks represent regions where selection has taken place and generally correspond to QTL positions for fruit weight or number of locules (source: M. Causse, according to Lin et al. (2014)). For a color version of this figure, see www.iste.co.uk/pelletier/genetics.zip*

This resequencing and gene fragment sequencing projects have provided several million ***SNPs*** for selection and enabled the creation of ***genotyping chips*** that have been used to characterize material from South American surveys or collections.

Blanca et al. (2015) studied the diversity of 1,000 accessions characterized by 7,000 SNPs. First, from accessions of *S. pimpinellifolium* in Ecuador, which are the closest to *S. l. cerasiforme* in Ecuador and had a similar habitat (humid forest), unlike accessions of *S. pimpinellifolium* from Peru, which grow in very arid conditions. Then, in Central America, a second stage of domestication, accompanied by a second bottleneck, gave rise to the first accessions of *S. l. cerasiforme*, then *S. lycopersicum*. Many cherry tomato varieties show an intermediate admixture structure between *S. pimpinellifolium* and *S. lycopersicum* (Figure 12.3).

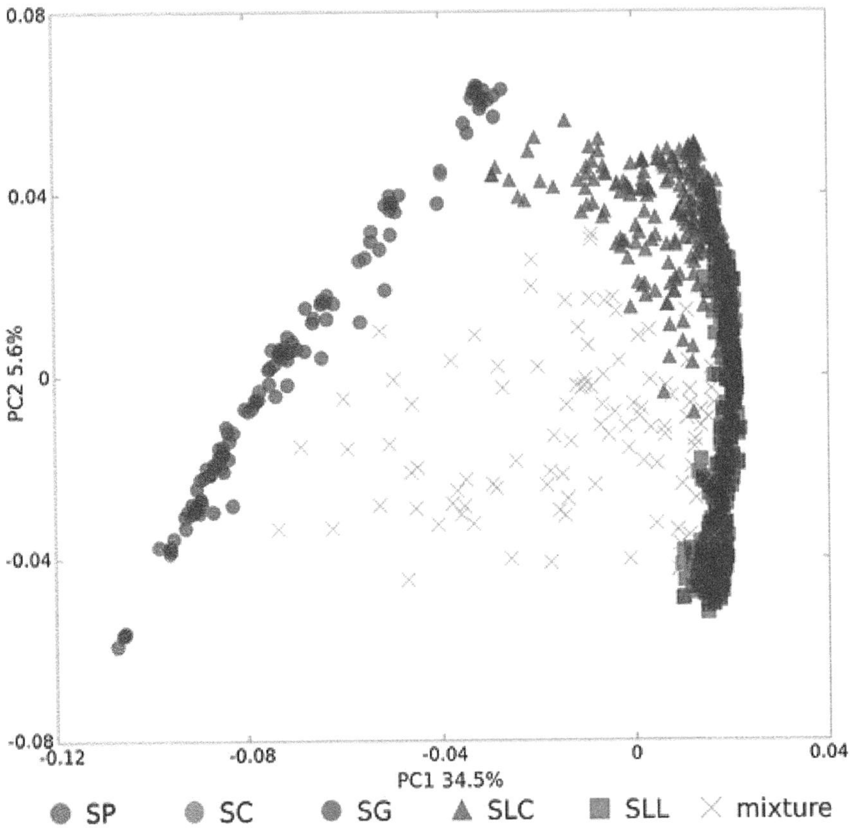

Figure 12.3. *Representation of the diversity of 1,000 tomato accessions of S. pimpinellifolium (SP), S. l. cerasiforme (SLC) or S. l. esculentum (SLL) (source: Blanca et al. (2015)). For a color version of this figure, see www.iste.co.uk/pelletier/genetics.zip*

12.3.3. *Some of the genes responsible for fruit shape diversity*

Recent discoveries of the molecular events shaping tomato diversity indicate that cultivated material is often more phenotypically diverse than wild genetic material, but does not necessarily show a similar pattern at the molecular level. Only a few genes are likely responsible for the high diversity of old varieties. Domestication and subsequent diversification of fruit types have led to great morphological diversity in the tomato fruit (with small to large, round, blocky, elongated or pear-shaped fruits, and colors ranging from red to green, white, black, pink, orange or yellow). In contrast, wild species that are closely related to the tomato bear small, round fruits that are red (*S. pimpinellifolium*), orange (*S. cheesmaniae*) or green (all other species), with low intraspecific phenotypic diversity.

Using molecular markers, the genetic control of fruit traits has been extensively dissected (Rothan et al. 2019). The first gene that controls variation in a quantitative trait locus (***QTL***) *fw2.2*, controlling variation in fruit weight, was cloned in 2000. The fruit shape diversity in cultivated genetic material could be explained, to a large extent, by four genes (Rodriguez et al. 2011) and a model of fruit shape evolution in the tomato was thus established. This model comprises four major mutations: LC and FAS, which increase the number of locules and fruit size; OVATE, responsible for the ovoid fruit shape; and SUN, which gives an elongated fruit shape or the "oxheart" shape when combined with LC and FAS. The distribution of mutations at the four genes, combined with morphological, geographical and historical data in a collection of various cultivated accessions, established that selection took place in distinct chronological and historical periods: LC appeared first, enabling the start of domestication, followed by OVATE, both of which were in *S. l. cerasiforme*, but in distinct populations. FAS appeared later in the presence of LC. The presence of these three mutations in Latin American genetic material suggests pre-Columbian mutations. Combined with *fw2.2*, they must have strongly contributed to the increase in fruit size during tomato domestication. In contrast, the SUN mutation is not carried by any Latin American material tested, suggesting that it appeared after domestication in Europe, probably in Italy (Figure 12.4).

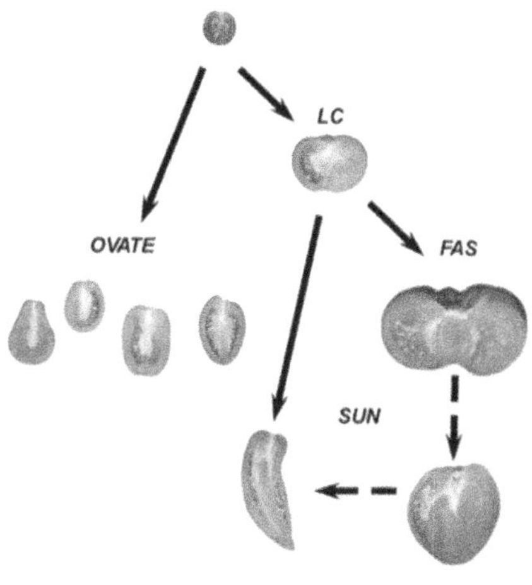

Figure 12.4. *Diagram of the appearance of fruit shape mutations from the wild type. LC, OVATE and FAS during the pre-Columbian period, SUN mutation during European post-domestication (source: M. Causse, according to Rodriguez et al. (2011)). For a color version of this figure, see www.iste.co.uk/pelletier/genetics.zip*

Other genes responsible for color diversity or insensitivity to the ***photoperiod*** (which allows the plant to flower, regardless of the length of day) have been identified (Rothan et al. 2019). Variations in the chemical composition of the fruit have also been brought about by domestication, notably in terms of taste (changes in sugar, acid and aroma content; Zhao et al. 2019).

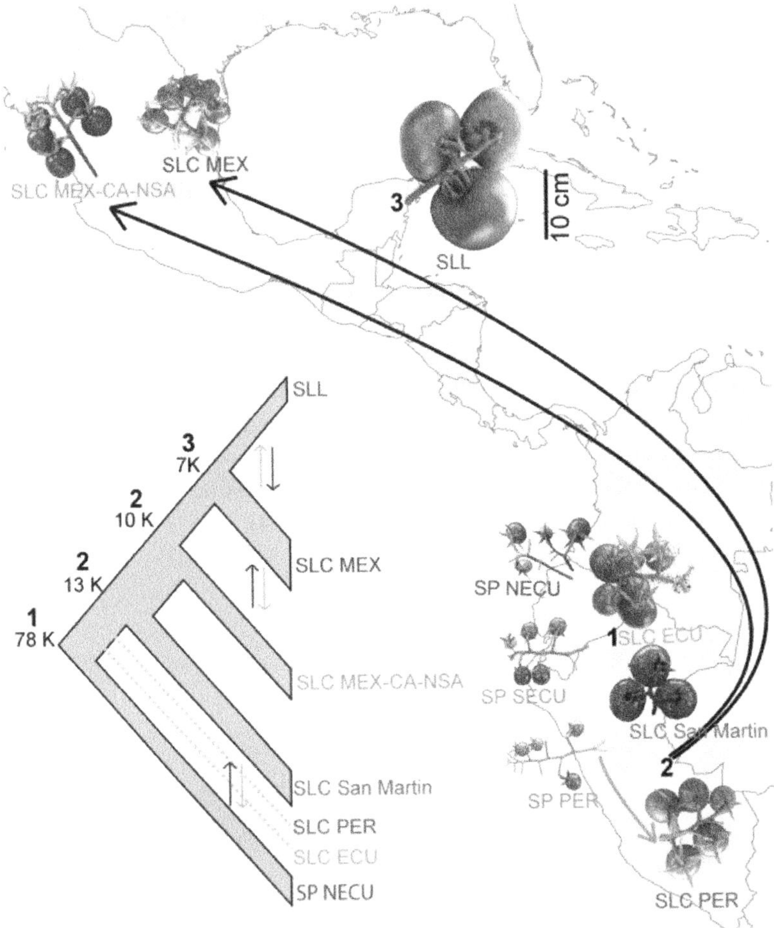

Figure 12.5. *Evolution of tomato domestication from S. pimpinellifolium (SP) to S. l. cerasiforme (SLC), then to S. l. lycopersicum (SLL) in Ecuador (ECU), Peru (PER), then Mexico (MEX) (source: M. Causse, according to Razifard et al. (2021)). For a color version of this figure, see www.iste.co.uk/pelletier/genetics.zip*

New sequencing techniques make it possible to resequence a large number of varieties, as well as to assemble certain genomes de novo. For example, the construction of a tomato ***pan-genome*** (based on the resequencing of 725 genomes and new assembly algorithms) showed that around 4,870 genes had not been detected in the reference genome, and that their presence varied according to variety (Gao et al. 2019). This particularly concerns many genes linked to stages of domestication, notably with the loss of genes passing from *S. pimpinellifolium* to *S. l. cerasiforme*, then to *S. lycopersicum*. These analyses have also enabled us to refine the diagram describing tomato domestication (Rafizard et al. 2020) and to date it to around 8,000–10,000 years BP (Figure 12.5). ***Transcriptome*** analyses have also identified domestication-induced changes in genome expression.

12.4. Post-domestication of the tomato after global expansion

12.4.1. *Introduction and distribution in Europe and worldwide*

The origin of the varieties that have been transferred to Europe is uncertain. Two hypotheses have been put forward concerning their origin. The first assumes that some seeds were transported to Europe from Peru sometime after the discovery of this region. The second hypothesis supports a Mexican origin. The most common hypothesis suggests that it was after the conquest of Mexico, which began in 1529, that the tomato was brought back to the Old World by the *conquistadors*. The existence of successive introductions from both origins has not been ruled out.

The first descriptions of the tomato appeared in the writings of naturalists from the 16th century, but these are brief and reflect its recent introduction to Europe. Around 1550, the Italians were the first Europeans to grow and consume tomatoes. A rapid migration to other countries is understandable, given that herbalists were always on the lookout for new and exotic medicinal plants. So, 25 years after its arrival in Italy, the tomato was introduced as a curiosity to botanical gardens in France, England, Spain and central Europe. Matthiolus, a 16th-century botanist, describes both eggplant and tomato in the first edition of Dioscorides. His 1554 definition of the tomato was translated into French in 1572 by Guillaume Rouillé (text translated into English): "Moreover, not long ago, we began to see another kind of flat eggplant, round like apples, divided into ribs like pompons, first green, then when ripe, yellow like gold on some plants and red on others. They are popularly known as *Pomi d'oro*, golden apples (the current common name for tomatoes in Italy). We eat them like Southerners (fried in oil): but they make you want to vomit".

The first published images of the tomato show a fairly large, very ribbed, red or yellow fruit (Figure 12.6). Later, Carl Linnaeus described the tomato in the same way, but also indicated that small-fruited plants (cherry tomatoes) had been cultivated for decorative purposes for several years. However, it was not until 1813 that Michel Félix Dunal described pear-shaped fruits, which he named *Lycopersicum piriforme*. He then described several types: *L. pomiferum*, *cerasiforme* and *humboldtii*.

The tomato was first preserved in herbalists' gardens before being disseminated throughout the Mediterranean region, where it remained little eaten for a long time. Its botanical proximity to mandrake led to it being used as a medicinal plant. It was even suspected of having aphrodisiac properties. Some authors began to question its nutritional qualities, while others continued to believe it was a poison. It was through Naples, a Spanish possession, that it entered Italian cuisine in the 16th century. The Italians passed it on to the people of Provence (via Genoa and Nice). It was not until the following generation that the people of Provence began to eat it. For a long time, the French believed that the tomato was poisonous and regarded it as an ornamental plant. It spread to northern Europe, but the fruit was only eaten in countries around the Mediterranean.

Figure 12.6. *An early depiction of a tomato plant (source: Matthiolus 1544). For a color version of this figure, see www.iste.co.uk/pelletier/genetics.zip*

The first French tomatoes were grown in Provence, notably in Châteaurenard (the "Plate de Châteaurenard" variety is still known today), Marseille and Perpignan. In 1790, the people of Marseilles introduced tomatoes to the people of Paris, to mark the first anniversary of the French Revolution. Later, the development of railroads enabled tomatoes from Provence to be marketed in northern France. One example is the "PLM" variety, named after the French railway company. More and more different types are appearing. Cookbooks mention the use of tomatoes in sauces and soups, or in recipes where the tomato is fried in oil or eaten raw with oil and vinegar (Daunay et al. 2007). This species was considered poisonous in Great Britain until the end of the 18th century, when the first vegetable varieties appeared. In the 19th century, tomato production underwent considerable development, thanks to advances in irrigation that increased both area and yield. This cultivation benefited from the crises suffered by other Mediterranean crops (grapevine, madder, olive, mulberry). It is also grown in the North African colonies. After the First World War, unsuccessful attempts were made to replace wine consumption with tomato juice.

At the same time, the tomato was brought to North America by British colonists, but its consumption spread slowly. Around 1810, Thomas Jefferson, President of the United States, is said to have convinced his fellow citizens that the fruit was not poisonous by eating it in front of a skeptical crowd. Some seed merchants included a few varieties (in the agronomic sense of the term) in their catalogs, first for ornamental use, then for horticulture. The consumption and production of this fruit then enjoyed unprecedented popularity, and seed companies began selecting certain cultivars for agronomic traits at the end of the 18th century. The recipe for *Tomato Ketchup* (a tomato and mushroom-based sauce originally derived from an Indonesian recipe), for example, was published during the American Civil War, in the southern United States, where food was scarce. Its production was subsequently industrialized by the Heinz company in 1876. The acidity of the fruit makes it particularly suitable for canning.

The autogamous nature of this species allows successive generations to resemble previous generations. Local cultivars are passed on to subsequent generations, earning them the name *Heirloom*. Growers sometimes select spontaneously occurring mutations that give the fruit or plant new characteristics, such as new fruit colors (black, white, dark purple, orange with green stripes, etc.), new fruit shapes (chili pepper shape), but above all, an increase in fruit size. It is thought, for example, that the elongated shape of the traditional southern Italian variety, *San Marzano*, comes from a mutation that only appeared in Italy after the species was transferred from Latin America. Waves of immigration to the United States have added to the diversity of cultivars that are already present.

12.4.2. *Tomato breeding in modern times*

Before 1920, tomato breeding was based on the selection of variants resulting from mutations or spontaneous cross-fertilization. Breeding began in the United States and later spread to other countries. It led to the development of cultivars adapted to different types of environment: the long hot season in Italy, the short field season in northern Europe, and greenhouses with low-light in winter in Holland. Numerous mutations have been discovered in the cultivated tomato species and characterized by genetic analysis (Rothan et al. 2019). For the most part, they correspond to single-gene mutations and to traits of agronomic interest. They involve plant architecture, fruit ripening and color, and male sterility.[4] For example, the *sp* (self-pruning) mutation confers a specific type of determinate growth and allows a limited development, which is much appreciated for the mechanical harvesting of industrial tomatoes. Since the 1970s, ***F1 hybrids*** have replaced pure line varieties. They are valued for their diverse agronomic characteristics, their stability and the possibility of combining several interesting traits in the same genotype. For example, it is possible to combine good fertility, early ripening and uniformity of plants and fruit, resistance to several pathogens and adaptability to different environments.

From its beginnings, tomato breeding has focused on traits such as yield, disease resistance and fruit quality. Tolerance to stresses induced by the plant's environment, whether biotic (fungal diseases, viral diseases and insect pests) or abiotic (unfavorable water, nutrition or temperature conditions), has been improved, notably with regard to the tomato's ability to develop in short days for production in winter greenhouses. Today's varieties, for example, set well in short days, whereas a few decades ago, the flowers of available varieties died in winter. Cold hardiness, tolerance to salinity and drought, and above all disease resistance, have been among the major objectives of breeding. To this end, South America's natural diversity has made an invaluable contribution, as we will see later.

Improving fruit quality has also been a breeding objective, but in different ways. In response to commercial demand, the external appearance of the fruit (shape, size and uniformity of shape and size) was improved. As a result, a range of cultivars with regular round fruit was obtained, and F1 hybrids with medium-sized fruit were created. Fruit color was another breeding objective, since it is a trait that influences consumer choice. A uniformly colored fruit with no green collar (top part of the fruit

4 Lack of viable pollen production in a normally bisexual plant, which can be determined by various nuclear or mitochondrial genes.

more intense green than the basal part before ripening) was obtained thanks to a recessive *u* gene (uniform ripening), identified in genetic resources and integrated into modern varieties from the 1970s onwards. Significant progress has also been made in obtaining intense red coloration. Resistance to bruising soon became one of the breeders' objectives, followed by firmness and shelf life. *Solanum pimpinellifolium* was an unexpected source of firmness, introduced as early as the 1960s in industrial varieties. Then, long-life varieties (varieties with a long shelf-life) were developed, to the detriment of taste quality, which was criticized by many consumers. Fruit ripening mutants (*rin* for ripening inhibitor, or *nor* for non-ripening) have a slowed or blocked ripening, with the fruit remaining firm for a very long time. These recessive mutations, used in breeding to increase fruit firmness and shelf life, do not allow for the development of aromas. Selection for taste has therefore become a new objective for some breeders. However, this selection has been limited as, until recently, there were no socioeconomic issues driving the development of this breeding. Today, consumers are offered a wide variety of tomato types to satisfy their multiple expectations.

12.4.3. *Using South American resources for multiple breeding objectives*

The 12 wild species related to the tomato can be crossed (more or less easily) with the cultivated tomato and have played a key role in plant breeding. Wild species have been the source of many pathogen resistance genes, and varieties resistant to one or more diseases have been developed. Adaptations to extreme environmental conditions (drought, salinity and cold) were obtained from crosses with related species adapted to these conditions. Charles Rick was one of the pioneers in the use of these species. From the 1950s onwards, he carried out several surveys in South America and built up one of the main collections of wild tomato relatives. His name was given to the C.M. Rick Tomato Genetics Resource Center at UC Davis, which has a large collection of wild and cultivated tomato resources. One of the first resistance genes introduced by selection in the United States, from the related species *S. pimpinellifolium*, gave resistance to *Fusarium oxysporum*. Subsequently, numerous wild accessions have brought new characteristics to cultivated tomatoes, giving new adaptations or resistance to other fungi, viruses and pests. It was also from *S. pimpinellifolium* that much firmer varieties were bred, facilitating mechanical harvesting of varieties intended for processing.

Crosses between wild and cultivated tomatoes can reveal alleles lost since domestication. Molecular markers have facilitated the use of these crosses in breeding. However, there are limits to their use. The difficulty of accurately introgressing alleles with a favorable effect without introducing unfavorable alleles, brought in by genetic linkage, remains a challenge. Genomic sequences provide clues for understanding the evolutionary history of the tomato ***clade*** and identifying the genes involved in fleshy fruit development. Schouten et al. (2019) have shown that the allelic diversity of tomato varieties stopped decreasing from the 1970s to 1980s, notably due to the introgression of resistance genes into new varieties from more distant wild sources. Advances in sequencing technologies are opening up new prospects in terms of genetic diversity management, particularly for the conservation and study of genetic resources. They are also opening up prospects for genome-wide selection through ***genomic selection*** (Duangjit et al. 2016).

In addition to the C. Rick collection, several tomato genetic resource collections were created in the 20th century to conserve tomato varieties, bringing together unique accessions, as well as accessions shared between institutes. These include the USDA[5] tomato collection, the N.I. Vavilov Research Institute of Plant Industry[6] (St Petersburg, Russia), the Asian Vegetable Research and Development Center (AVRDC) in Taiwan, the Leibniz Institute of Plant Genetics and Crop Plant Research[7] (Gatersleben, Germany), the Estación Experimental La Mayora (Algarrobo Costa, Málaga, Spain) and the tomato genetic resources collection of the INRAE Centre for Vegetable Germplasm in Avignon.[8] The European Cooperative Programme for Plant Genetic Resources Tomato Database[9] (Wageningen, Netherlands) is a database that has been developed to bring together all the information contained in European collections.

While ex situ genetic material conservation is well developed and will benefit from developments in genomics, in situ conservation of tomato and its wild relatives is becoming critical due to major ecological changes in the area of origin. In situ conservation efforts and participatory approaches could be very useful in maintaining the adaptive potential of tomato genetic resources. For example, it has been proposed to use accessions of *S. cheesmaniae* that are currently stored in genetic material banks to re-establish certain extinct populations in the Galapagos Islands. This could help avoid the current paradox: the more we learn about tomato diversity and its evolutionary history, the less these genetic resources are available in the wild.

5 Available at: http://www.ars-grin.gov/npgs/searchgrin.html.

6 Available at: http://www.vir.nw.ru/data/dbf.htm.

7 Available at: http://pgrc.ipk-gatersleben.de.

8 Available at: https://eng-gafl.paca.hub.inrae.fr/vegetable-germplasm-centre.

9 Available at: https://www.ecpgr.cgiar.org/working-groups/solanaceae.

12.5. Conclusion

The tomato has become an essential part of our diets worldwide. It is a symbol of the Mediterranean diet, so much so that it is celebrated in many places.[10] Despite this importance, few details are known about its ancient history. By the time the tomato arrived at the Spanish court in the 16th century, it had been cultivated for several centuries and was already widely used by the Aztecs. However, we do not accurately know when and where it was domesticated.

Its history across the continents has led to a sharp reduction in genetic diversity, despite the boom in morphological variation for which it is known. This reduction in diversity is due to three successive bottlenecks: (i) domestication, which undoubtedly began unknowingly in the Andean valleys before migration to Mexico; (ii) the passage of a few cultivars to Europe; (iii) modern breeding in Europe and the United States, which has further reduced the pool of progenitors used in breeding programs. Fortunately, related species from South America have been used to restore some of the original diversity and discover new characteristics that have enabled this species to adapt to a wide range of conditions. Genetic resource collections of old accessions are essential relics for the improvement of tomorrow's varieties.

12.6. References

Bai, Y. and Lindhout, P. (2007). Domestication and breeding of tomatoes: What have we gained and what can we gain in the future? *Annals of Botany*, 100, 1085–1094.

Blanca, J., Montero-Pau, J., Sauvage, C., Bauchet, G., Illa, E., Diez, M.J., Francis, D., Causse, M., van der Knaap, E., Cañizares, J. (2015). Genomic variation in the tomato, from wild ancestors to contemporary breeding accessions. *BMC Genomics*, 16, 257. doi: 10.1186/s12864-015-1444-1.

Daunay, M.C., Laterrot, H., Janick, J. (2007). Iconography and history of Solanaceae: Antiquity to the 17th century. *Horticulture Review*, 34, 1–119.

Duangjit, J., Causse, M., Sauvage, C. (2016). Efficiency of genomic selection for tomato fruit quality. *Molecular Breeding*. doi: 10.1007/s11032-016-0453-3.

Lin, T.G., Zhu, J., Zhang, X., Xu, Q., Yu, Z., Zheng, Z., Zhang, Y., Lun, S., Li, X., Wang, Z. et al. (2014). Genomic analyses provide insights into the history of tomato breeding. *Nature Genetics*, 46, 1220–1226. doi: 10.1038/ng.3117.

---

10 One example is La Tomatina, in Buñol, Spain, where every year at the end of August, several tons of overripe tomatoes are used as projectiles in a giant battle between ever-growing numbers of people.

Peralta, E., Spooner, D., Knapp, S. (2008). The taxonomy of tomatoes: A revision of wild tomatoes (*Solanum* section *Lycopersicon*) and their outgroup relatives in sections *Juglandifolium* and *Lycopersicoides*. *Systematic Botany Monography*, 84, 1–186.

Razifard, H., Ramos, A., Della Valle, A.L., Bodary, C., Goetz, E., Manser, E.J., Li, X., Zhang, L., Visa, S., Tieman, D. et al. (2020). Genomic evidence for complex domestication history of the cultivated tomato in Latin America. *Molecular Biology and Evolution*, 37(4), 1118–1132. doi: 10.1093/molbev/msz297.

Rick, C.M. (1976). Tomato *Lycopersicon esculentum* (Solanaceae). *Evolution of Crop Plants*, 1, 268–273.

Rodriguez, G.R., Munos, S., Anderson, C., Sim, S.C., Michel, A., Causse, M., McSpadden Gardener, B.B., Francis, D., van der Knaap, E. (2011). Distribution of SUN, OVATE, LC, and FAS in the tomato germplasm and the relationship to fruit shape diversity. *Plant Physiology*, 156(1), 275–285.

Rothan, C., Diouf, I., Causse, M. (2019). Trait discovery and editing in tomato. A review. *The Plant Journal*, 97, 73–90. doi: 10.1111/tpj.14152.

Schouten, H.J., Tikunov, Y., Verkerke, W., Finkers, R., Bovy, A., Bai, Y., Visser, R. (2019). Breeding has increased the diversity of cultivated tomato in the Netherlands. *Frontiers in Plant Science*, 10, 1606. doi: 10.1606,10.3389/fpls.2019.01606.

TGC, The Tomato Genome Consortium (2012). The tomato genome sequence provides insights into fleshy fruit evolution. *Nature*, 485, 635–641.

Zhao, J.T., Sauvage, C., Zhao, J., Bitton, F., Bauchet, G., Liu, D., Huang, S., Tieman, D., Klee, H.J., Causse, M. (2019). Meta-analysis of genome-wide association studies provide new insights into genetic control of tomato flavour. *Nature Communications*, 10, 1534. doi: 10.1038/s41467-019-09462-w.

13

Mutagenesis and Accelerated Domestication

Georges PELLETIER[1,2]
[1]*Académie des sciences, Paris, France*
[2]*Académie d'agriculture de France, Paris, France*

While the number of higher plant species in the world is estimated at between 300,000 and 500,000, around 7,000 of them are thought to have been consumed in the wild for millennia. Then, humanity's choice of some 2,500 species, subject to a variety of consumption patterns and cultivation practices in different environments, led to a slow selection process that resulted in the complete domestication of just 10% of these species. Today, some 5–10 species, including wheat, corn, rice, palm oil and potato, make up 95% of mankind's caloric intake. This concentration makes food production vulnerable to climate change, as well as the possible emergence of pests and diseases. Furthermore, the genetic diversity present in cultivated forms has been considerably reduced.

Comparative studies of wild and domesticated forms have revealed several mutations responsible for domesticated phenotypes, sometimes shared by different species (Kumar et al. 2021). This knowledge makes it possible to imagine, on the one hand, the de novo domestication of cultivated species, starting from populations of the wild species that were not subjected to domestication, in order to explore and exploit other genetic diversity in crops; or, on the other hand, the domestication of a wild species that has not (or imperfectly) been domesticated, so as to adapt it to agricultural production (Oster-berg et al. 2017). These prospects are now becoming

Genetics of Domestications,
coordinated by Georges PELLETIER.

realistic with the development of mutagenesis techniques, and in particular the targeted modification of genes by genome ***editing*** (Fernie and Yan 2019).

13.1. Random mutagenesis and neo-domestication

13.1.1. *Random mutagenesis*

As we have seen in previous chapters, the process of plant domestication, which sometimes only involves a few genes with major effects, has been slow, as it depends on the frequency of occurrence of a favorable spontaneous mutation, as well as the selective pressure applied to it. However, the order of magnitude of spontaneous mutation frequency is fairly constant in plants, corresponding to around one DNA base pair in 100 million. Under these conditions, the selection of a favorable mutation requires a considerable number of individuals that are compatible in the long term with even primitive agricultural production, but incompatible with the activity of a varietal selection laboratory. Therefore, in the first half of the 20th century, the mutagenic effects of certain types of radiation (X-rays, gamma rays, ultraviolet rays, etc.) were researched and discovered; and in the second half of the 20th century, those of certain substances (sodium azide, ethyl methane sulfonate, etc.). For almost a century, mutagenesis through the processing of seeds, buds, pollen or cells grown in vitro has thus become a common tool in plant breeding, with the mutants produced being added to natural mutants, and both types of mutants being used indiscriminately in the creation of plant varieties. In 2020, among 170 species, over 3,360 varieties were known to have undergone mutagenic treatments, mainly by radiation, without counting all those derived from cross-breeding, which are not listed.

The use of mutagenic treatments, whether chemical or physical, considerably increases the frequency of spontaneous mutations by a factor of around 1,000, proportionally reducing the number of plants to be observed and selected to obtain the desired mutation. However, the total number of mutations in this case is still around 1,000 times smaller than the number of differences between the DNAs of two unrelated individuals of the same species. Different types of mutation can be obtained, depending on the mutagenic treatment used, such as a chemical agent, irradiation or stress in in vitro cell culture. Together, these mutagens can reproduce the different types of spontaneous mutations, whether punctual (change of a base pair) or structural (deletion, nucleotide insertion, translocation). Induced mutations are similar in nature to spontaneous mutations, which are themselves highly diverse, as demonstrated by the genetic variability between ***accessions*** of the same species that results from their accumulation over generations.

13.1.2. *The example of Vigna stipulacea*

An example of de novo domestication by random mutagenesis involves the semi-wild legume (*Fabaceae*) species *Vigna stipulacea*, native to the Indian peninsula and present in South Asia. This species is used as fodder, and its nutritious seeds are occasionally eaten by humans. It is fast-growing with a short vegetative phase and has the advantage of being resistant to many crop diseases and pests. However, its pods easily release their seeds (dehiscence), and the latter enter dormancy for varying lengths of time, which limits the possibilities of uniform harvesting and sowing, and, ultimately, cultivation and consumption.

From 12,000 seeds treated with the chemical mutagen ethyl methane sulfonate, 3,000 plants were bred and multiplied to produce 18,000 plants in the next generation. These included a first-generation plant with non-dehiscent pods and three second-generation lines with different dormancy reductions (Takahashi et al. 2019). The corresponding mutations at different *loci* each concern a single gene. As the genome of this species has also been sequenced, the identification of the genes involved will make it easier to introduce them by cross-breeding in different accessions of the species; also, it will make it easier to improve the same traits in other wild or cultivated legumes. Random mutagenesis and the selection of phenotypes facilitating cultivation are applicable to most species, and although they require the observation of numerous plants, they remain competitive with targeted mutagenesis methods, which we will illustrate in species where notable cell culture and whole-plant in vitro regeneration techniques have been mastered.

13.2. Genome editing and domestication

13.2.1. *Genome editing*

Recent years have seen the development of targeted genome modification methods. The latter are targeted because a given genome sequence is recognized by the molecular players (proteins and RNA) involved.

Initially, these players include nucleases that cause cuts in the DNA double chain, earning them the name "molecular scissors", either TALEN (Transcription Activator-Like Effector Nucleases) or CRISPR Cas9 (Clustered Regularly Interspaced Short Palindromic Repeats) (Christian et al. 2010; Jinek et al. 2012). This clean cut of the two DNA strands enables three types of intervention on this genome (see Figure 13.1). In the TALEN method, the cut site is recognized by the DNA-binding domains of chimeric proteins that are associated with a nuclease that carries out the cut. In the CRISPR Cas method, a nuclease (Cas9 or others since discovered) is associated with a guide RNA which, complementary to one of the two DNA strands of the target gene, positions this nuclease, which then performs a DNA cut at a determined site. In the absence of a template (Figure 13.1, SDN1), the DNA will be repaired randomly, generally resulting in inactivation of the target gene.

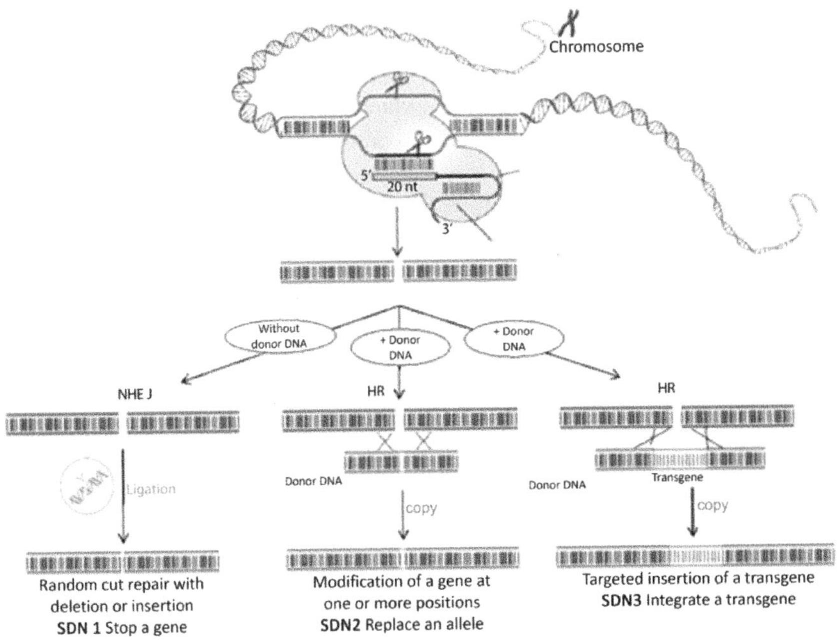

Figure 13.1. *Illustration of the molecular scissors (CRISPR Cas9), which cuts both strands of the DNA molecule at a given site, and three types of genome-editing intervention, targeting a single site in a genome: SDN1, SDN2, SDN3, for site-directed nuclease 1, 2 and 3 (authorization: F. Nogué, INRAe Versailles). For a color version of this figure, see www.iste.co.uk/pelletier/genetics.zip*

COMMENT ON FIGURE 13.1.– *SDN1: after cutting both strands of the DNA molecule, DNA repair systems produce a nucleotide insertion or a micro-deletion, which may result in the inactivation of the desired gene. This type of intervention is known as SDN1. SDN2: cut repair is performed in the presence of a DNA template homologous to the cut site region, but differing in one or more mutations. By homologous recombination at the cut site, this template introduces these mutations into the genome. This type of intervention is known as SDN2 and mimics the replacement of one allele by another through cross-breeding and selection.*

SDN3: the cut is made in the presence of a DNA template homologous to the region of the cut site, but containing a transgene, a gene that may or may not be foreign to the species' genome, which will be inserted by homologous recombination at the cut site. This type of intervention is known as SDN3 or targeted transgenesis.

Other rewriting methods derived from the properties of the CRISPR system have since emerged to achieve increasingly precise modifications. Hybrid proteins between the Cas9 nuclease, whose nuclease function has been inactivated, and a cytidine deaminase or an adenosine deaminase, make it possible to chemically transform a cytidine (C) into a uridine (U), or an adenine (A) into a guanine (G); in other words, to target the transition from a GC to an AT base pair or vice versa. The combination of an inactivated Cas9 reverse transcriptase and the extension of the guide RNA by a sequence that is destined to be converted into DNA enables the direct rewriting of a longer genome sequence (prime editing). These last methods, which still have relatively limited success rates, are currently under intense research to improve them (Liu et al. 2023).

13.2.2. *De novo tomato domestication*

The domestication and selection of cultivated plants to feed the world has been achieved by exploiting a reduced share of the genetic diversity of ancestral and neighboring species. We can speak of a bottleneck in genetic diversity in the early stages of domestication. Over the last hundred years or so, breeders of new varieties have resorted to hybridization with wild species, so as to introduce traits that improve harvest quality, disease resistance or adaptation to unfavorable climatic conditions. There are many examples of this in wheat, potato and tomato, to name but a few major species. The task is a complex one, as it involves sorting out the cumulative domestication and improvement traits from the undesirable wild traits in the progeny.

Another approach is to start from different origins of the wild ancestral species and accelerate domestication of these plants by modifying genes known to have undergone domestication mutations using genome editing tools.

One example is the tomato (*Solanum lycopersicum*) and its ancestor *Solanum pimpinellifolium*, where we find accessions that are capable of adapting to diverse and sometimes hostile environments, such as deserts, as well as others that are resistant to the black spot disease caused by bacteria of the genus *Xanthomonas*, which can be devastating under certain conditions, and those that are resistant to salty soils, etc.

The domestication genes that were targeted by two independent teams (Li et al. 2018; Zsögön et al. 2017) were first:

– the *SP* (Self-Pruning) gene from the family of flowering inhibitors, whose inactivation transforms indeterminate growth into determinate growth[1];

– the *SP5G* gene, whose inactivation causes insensitivity to daylength, enabling long-day cultivation at higher latitudes than in the tropical zone of origin. The combination of the two mutations also allows for earlier flowering and harvesting.

Then, to increase vitamin C concentration, the expression level of the gene encoding an enzyme (*GGP1*) in the biosynthesis pathway was increased by mutation of its *promoter*.

The shape of the fruit, which is cylindrical at the outset, has been modified by inactivation of the OVATE gene, resulting in a more oval shape that reduces the risk of bursting at maturity, as observed with wild fruit.

Fruit size and the number of locules were increased by inactivation of the *CLAVATA 3* gene, almost tripling their weight. The number of fruits was increased by inactivation of the *MULTIFLORA* gene. Lastly, lycopene content, a powerful antioxidant, was increased more than fivefold by inactivation of the lycopene β-cyclase gene, the enzyme that converts it to β-carotene in wild plants.

Directed mutagenesis of the genome of wild ancestors, by carrying out a partial but rapid domestication that results in plants that can be compared to those of today's varieties (even if the fruit remains of moderate size), can thus be used to more easily introduce lost but desirable characteristics into the selection of cultivated plants.

13.2.3. *Domestication of Physalis pruinosa*

Physalis pruinosa, also known as *ground cherry* or *husk cherry* (Figure 13.2), is an annual plant belonging to the *Solanaceae* family, which has retained many of the characteristics of a wild plant that limit its cultivation; this is restricted to Central and South America, and its consumption: an exacerbated bushy growth pattern in all directions, the result of indefinite (indeterminate) growth of stems and branches, and small fruits that fall spontaneously to the ground. These are all characteristics found

1 Growth is indeterminate when new structures (branches, flowers, fruit) are formed indefinitely. Growth is determinate when it ceases as soon as a given structure (such as a certain number of flowers) is fully formed.

in tomatoes' wild ancestor, *Solanum pimpinellifolium*. The evolutionary proximity between the two species led the authors to consider the major tomato domestication genes as candidates for modification, using genome editing tools for the corresponding *Physalis pruinosa* genes. To this end, in vitro culture methods and the transfer of molecular genome editing tools have been developed, and the development of genomic analysis of this species has made it possible to identify almost 13,000 ***orthologous*** tomato genes.

Figure 13.2. *Physalis pruinosa*[2]. *For a color version of this figure, see www.iste.co.uk/pelletier/genetics.zip*

In cultivated tomatoes and peppers, varieties with determinate growth (which stops when certain organs are formed or when a certain size is reached) are due to the inactivation of one of the members of the *SP* gene family, which regulates the transition from the vegetative state to the reproductive state. In *Physalis pruinosa*, inactivation of the *SP5G* gene causes a more compact, bushy growth habit, resulting in an increase in the number of flowers and fruits along each shoot. By inactivating the ortholog of the tomato *AGO7* gene, the size of leaves and floral organs is reduced; by inactivating the ortholog of the *CLV3* gene, an increase in the size of

2 Available at: https://www.ncbi.nlm.nih.gov/Taxonomy/taxi/images/14956.

fruits with a greater number of lodges is obtained; lastly, by inactivating the ortholog of the *JOINTLESS* gene, the fruits remain attached and allow harvesting at maturity. All these modifications contribute to the possibility of growing this species more extensively in the field (Lemmon et al. 2018).

13.2.4. *Domestication of Oryza alta*

Today, rice is the staple food of more than half the world's population, and it is also the second most important cereal in the direct human diet, after wheat. Its domestication, which has already been mentioned (see Chapter 10), has focused on a number of important agronomic characteristics, such as erect plant architecture, non-dispersal of grains, panicle morphology, ear length, absence of awns, grain quality and size, and envelope color.

This ***diploid*** rice, *Oryza sativa*, has a relatively simple genome that was among the first to be sequenced. Since then, thousands of genomes of various cultivated and wild varieties have been described. The function of well over a hundred genes that affect agronomic and domestication traits has also been identified.

There are 27 species in the genus *Oryza*, whose genomes are named by letters of the alphabet. *Oryza sativa* (AA) carries the A genome and five other diploid genomes are known: BB, CC, EE, FF and GG. In addition, there are five ***allotetraploid*** genomes (BBCC, CCDD, HHJJ, HHKK and KKLL) (Wing et al. 2018). CCDD species are derived from the hybridization of a CC species (*Oryza officinalis* or *Oryza rhizomatis*) and a now-extinct DD species closely related to *Oryza australiensis*. Those from South America have many interesting traits for agricultural production, such as a much greater biomass and resistance to disease, insects and climatic variations. The cultivation of ***polyploid*** rice has been proposed as a way of benefiting from these advantages, but has so far come up against a number of difficulties, the main one being limited genetic resources, particularly for certain domestication ***alleles*** in these species.

Among polyploid species, *O. alta* was chosen (Figure 13.3), with authors (Yu et al. 2021) targeting certain orthologous genes of *O. sativa* rice domestication genes by genome editing, which involved the development of the genomics of this ***tetraploid*** wild species (sequencing, identification of genome elements – coding sequences, ***transposons***, repeated sequences, ***microRNAs***, etc. – and their distribution between the C and D sub-genomes and efficient in vitro transformation techniques). To reduce spontaneous grain dispersal, *qSH1* orthologs (C and D genomes) were inactivated (SDN1-type intervention; see Figure 13.1) and *An-1* orthologs (C and D) were inactivated to shorten or eliminate awns. Similarly, inactivation of SD1 orthologs can reduce plant height (*O. alta* can exceed 2 m), following the example of the *sd1* mutation adopted in Green Revolution rice, which improves resistance to lodging, increases harvest index (ratio of grain biomass to total biomass) and thus grain yield. The *GS3* gene has been inactivated to improve grain length and size, as have the *Dhd7* and *DTH7* genes to advance the heading date from 150 days to just 80 days at higher latitudes than in its original tropical range, thus extending its cultivation areas. The sequence of the *IPA* (ideal plant architecture) gene was modified by base editing to increase its expression level, which is controlled by a microRNA. Under these conditions, the stems have a larger diameter, fuller panicles and, ultimately, higher yields.

a) b)

Figure 13.3. *Allotetraploid wild rice Oryza alta (a) and a line with a more compact morphology, where the OaSD1-CC and 0aSD1-DD genes are mutated by editing (b). Bars 35 cm. Figures generously provided by Dr. Hong Yu. For a color version of this figure, see www.iste.co.uk/pelletier/genetics.zip*

13.2.5. *Other candidate species for accelerated domestication*

In principle, a host of species could be candidates for accelerated domestication to adapt them to a given production, including the parent species of the current cultivated species, so as to integrate previously unused genetic variability into variety creation.

Crop adaptation to climate change (Razzaq et al. 2021) could include new species for oil production (*Helianthus deserticola*), protein production (*Vigna marina*) and starch production (*Panicum turgidum*), which are adapted to conditions of drought and high temperature. On the other hand, in regions where the climate is becoming warmer and more humid, starch, protein and oil production could use *Echinochloa stagnina*, *Centrosema molle* and *Gossypium tomentosum*, respectively.

We can also think of new crops of species to meet the new needs of the bio-economy. These include, for example, third-generation bioenergy crops, which need to combine adaptation to climatic challenges with productivity and versatility in the products they produce.

In this context, the wild species *Silphium perfoliatum* is a candidate for the production of biomass, quality oil and fodder by acquiring characteristics of the neighboring species *S. integrifolium*, which produces high biomass, with drought tolerance. The *Sida hermaphrodita* species could be used for fuel, fiber production, biogas and high-quality fodder. *Thinopyrum intermedium* is a perennial cereal that can be used as fodder; however, its grain yield, which is clearly inadequate, needs to be brought up to the level of annual cereals before it can be adopted for agricultural production (Wever et al. 2020).

13.3. Limits and constraints of neo-domestication

The development of these methods is subject to obstacles of all kinds.

13.3.1. *In practical terms*

The use of random mutagenesis, based on phenotypic selection, hardly requires any prior knowledge or expertise. On the other hand, the number of plants to be subjected to selection, the need to repeat the selection for each key trait, and the crosses required to accumulate the different mutated alleles in a single genotype, all make for restrictive experimental set-ups.

The use of genome editing techniques is applicable, provided that we first have a thorough knowledge of the genomes of candidate wild species and neighboring species that can be used as models, in order to determine the domestication genes and their orthologs in the target species. This first step is becoming increasingly accessible, with the development of modern genome sequencing methods and access to model species databases.

The second condition, which is currently more limiting, is the possibility of introducing the molecular factors required for editing into a plant cell, which are then capable of regenerating an entire plant. In reality, few plant species have been subjected to in vitro culture techniques, and for those that have, the possibilities are usually limited to certain genotypes. These molecular players can be introduced in the form of transgenes, preferably expressed transiently. If they are integrated into the genome (transgenesis), they must be removed by segregation and selection in the progeny. This is not feasible if you want to preserve the characteristics of a variety reproduced by vegetative propagation, as is the case with, for example, grapes, apples or potatoes. Molecular complexes, such as "Cas9-guide RNA", can be pre-assembled in protoplasts[3] derived from somatic tissues (tobacco, lettuce, rice, grapevine, apple and potato), in embryonic cells by bombardment (corn and wheat), or in zygotes after in vitro fertilization (rice).

The domestication of wild plants is only complete, in the sense of their integration into agricultural practices in addition to (or in competition with) current cultivated plants, if dozens of genes have been modified and assessed in the field. The time needed to create a new high-performance crop can therefore take several years, or even decades, depending on the number of genes to be modified and the duration (annual, perennial) of the species' reproductive cycle. In addition to this, there needs to be acceptance by growers, which depends on the emergence of new markets and are themselves subject to the reactions of industrial chains or consumers, as well as to regulations facilitating or hindering the cultivation of edited plants.

13.3.2. *In terms of health and environmental safety*

Clearly, not all wild species, including those related to cultivated species, are devoid of compounds that are toxic to humans or animals. This must therefore be considered when selecting species for domestication, and they must be eliminated by mutagenesis when necessary.

Another aspect often put forward concerns the consequences of the genome modification techniques themselves. In the case of genome modifications without

3 Plant cells stripped of their pectocellulosic wall by enzymatic treatment, which can be cultivated in vitro to obtain an entire plant in certain species.

the "addition of foreign DNA", some focus on the potential dangers of inducing mutations elsewhere in the genome than in the targeted sequences. These so-called "off-target" mutations are relatively rare in plants in most of the cases studied, and their number is incommensurate with the thousands of mutations induced by mutagenesis methods developed in breeding for almost a century, that are rightly considered safe (Sturme et al. 2022).

13.3.3. *In legal terms*

The neo-domestication of wild species confined to regions owned or managed by indigenous communities for centuries is subject to the provisions of international agreements: the Convention on Biological Diversity, and more specifically, the Nagoya Protocol, which stems from it.[4]

The objective of the latter "is the fair and equitable sharing of benefits arising from the utilization of genetic resources, including by appropriate access to genetic resources and by appropriate transfer of relevant technologies, taking into account all rights over those resources and to technologies, and by appropriate funding, thereby contributing to the conservation and sustainable use of biodiversity". This applies to "1) genetic resources within the scope of the Convention on Biological Diversity[5]; (2) benefits arising from the utilization of such genetic resources; (3) traditional knowledge associated with genetic resources; (4) benefits arising from the utilization of such knowledge". Any neo-domestication program based on exotic genetic resources is therefore subject to the provisions of these international agreements.

13.3.4. *In regulatory terms*

Accelerated domestication generally requires the mutation of several genes to inactivate them or modify their regulation. It can therefore use SDN1 or SDN2 genome editing techniques (see Figure 13.1). The regulations applying to the products of these targeted modifications vary enormously (for the time being?) from country to country. For example, India considers that there is no scientific reason to regulate SDN1 and SDN2 products as GMOs. Conversely, New Zealand prohibits the cultivation of plants derived from targeted mutagenesis, classifying them as GMOs. The first products derived from genome editing are already authorized in

4 Available at: https://biodiv.mnhn.fr/fr/documents/protocole-de-nagoya.

5 Available at: www.cbd.in.

certain regions of the world, such as the Americas. Such cacophony will pose considerable problems for the international exchange of seeds and products derived from these varieties. Europe is seeking to reconcile its future regulations with those for genetically modified organisms, at the risk of completely blocking their development. The European directives and regulations drawn up over 30 years ago need to be updated in line with the progress of scientific knowledge, rather than the ideologies that still prevail; otherwise, Europe will miss out on the opportunities offered by genome editing methods.

13.4. Conclusion

Many species have been cultivated in the past, and some have been gradually abandoned in favor of species better suited to agricultural production. The selection of spontaneous genetic modifications by our ancestors since Neolithic times has enabled the domestication of these species and their gradual improvement. These "novelties" were clearly appreciated, used and happily passed on to subsequent generations.

Until a few years ago, the molecular nature of these modifications was completely unknown. Advances in genomics have made it possible to accumulate a considerable amount of data on the function of the genes involved in the domestication process and the changes to their DNA sequences.

The prospect of re-domesticating wild plants, or those that have become semi-wild after being cultivated, is becoming completely realistic with advances in our knowledge of domestication genes and in targeted genome modification techniques. It will only become a reality if a certain number of technical, legal and regulatory obstacles are overcome, particularly mistrust, sometimes even leading to rejection, which is paradoxical in our modern societies, with regard to the genetic modification of plants carried out by humans as soon as they have knowledge of the DNA sequences involved!

13.5. References

Christian, M., Cermak, T., Doyle, E.L., Schmidt, C., Zhang, F., Hummel, A., Bogdanove, A.J., Voytas, D.F. (2010). Targeting DNA double-strand breaks with TAL effector nucleases. *Genetics*, 186(2), 757–761. doi: 10.1534/genetics.110.120717.

Fernie, A.R. and Yan, J. (2019). De novo domestication: An alternative route toward new crops for the future. *Mol. Plant.*, 12, 615–631. doi: 10.1016/j.molp.2019.03.016.

Jinek, M., Chylinski, K., Fonfara, I., Hauer, M., Doudna, J.A., Charpentier, E., (2012). A programmable dual-RNA-guided DNA endonuclease in adaptive bacterial immunity. *Science*, 337, 816–821. doi: 10.1126/science.1225829.

Kumar, R., Sharma, V., Suresh, S., Ramrao, D.P., Veershetty, A., Kumar, S., Priscilla, K., Hangargi, B., Narasanna, R., Pandey, M.K. et al. (2021). Understanding omics driven plant improvement and de novo crop domestication: Some examples. *Front. Genet.*, 12, 637141. doi: 10.3389/fgene.2021.637141.

Lemmon, Z.H., Reem, N.T., Dalrymple, J., Soyk, S., Kerry, E., Swartwood, K.E., Rodriguez-Leal, D., Van Eck, J., Lippman, Z.B. (2018). Rapid improvement of domestication traits in an orphan crop by genome editing. *Nature Plants*, 4, 766–770. doi: 10.1038/s41477-018-0259-x.

Li, T., Yang, X., Yu, Y., Si, X., Zhai, X., Zhang, H., Dong, W., Gao, C., Xu, C. (2018). Domestication of wild tomato is accelerated by genome editing. *Nat. Biotechnol.*, 36, 1160–1163. doi: 10.1038/nbt.4273.

Liu, T., Zhang, X., Li, K., Yao, Q., Zhong, D., Deng, Q., Lu, Y. (2023). Large-scale genome editing in plants: Approaches, applications, and future perspectives. *Current Opinion in Biotechnology*, 79, 102875.

Østerberg, J.T., Xiang, W., Olsen, L.I., Edenbrandt, A.K., Vedel, S.E., Christiansen, A., Landes, X., Andersen, M.M., Pagh, P., Sandøe, P. et al. (2017). Accelerating the domestication of new crops: Feasibility and approaches. *Trends in Plant Science*, 22(5), 373–384. doi: 10.1016/j.tplants.2017.01.004.

Razzaq, A., Wani, S.H., Saleem, F., Yu, M., Zhou, M., Shabala, S. (2021). Rewilding crops for climate resilience: Economic analysis and de novo domestication strategies. *Journal of Experimental Botany*, 72(18), 6123–6139.

Sturme, M.H., van der Berg, J.P., Bouwman, L.M., De Schrijver, A., de Maagd, R.A., Kleter, G.A., Battaglia-de Wilde, E. (2022). Occurrence and nature of off-target modifications by CRISPR-Cas genome editing in plants. *ACS Agricultural Science & Technology*, 2(2), 192–201.

Takahashi, Y., Sakai, H., Yoshitsu, Y., Muto, C., Anai, T., Pandiyan, M., Senthil, N., Tomooka, N., Naito, K. (2019). Domesticating *Vigna Stipulacea*: A potential legume crop with broad resistance to biotic stresses. *Front. Plant Sci.*, 10, 1607. doi: 10.3389/fpls.2019.01607.

Tan, L.B. and Sun, C.Q. (2021). Rapid domestication of wild allotetraploid rice: Starting a new era of human agricultural civilization. *Chinese Bulletin of Botany*, 56(2), 134–137. doi: 10.11983/CBB21022.

Wever, C., Tassel, D.L.V., Pude, R. (2020). Third-generation biomass crops in the new era of De novo domestication. *Agronomy*, 10(9), 1322.

Wing, R.A., Purugganan, M.D., Zhang, Q. (2018). The rice genome revolution: From an ancient grain to Green Super Rice. *Nat. Rev. Genet.*, 19, 505–517.

Yu, H., Lin, T., Meng, X., Du, H., Zhang, J., Liu, G., Chen, M., Jing, Y., Kou, L., Li, X. et al. (2021). A route to de novo domestication of wild allotetraploid rice. *Cell*, 184, 1156–1170. doi: 10.1016/j.cell.2021.01.013.

Zsögön, A., Cermak, T., Voytas, D., Pereira Peres, L.E. (2017). Genome editing as a tool to achieve the crop ideotype and de novo domestication of wild relatives: Case study in tomato. *Plant Science*, 256, 120–130. doi: 10.1016/j.plantsci.2016.12.012.

Glossary

Accession: This includes element of plant genetic resource conservation; it can be a variety or a population.

Allele: This refers to a variant form of a locus, which can lead to a change in the function or activity level of said locus if it corresponds to a gene. The allele can be identified by sequencing.

Ancient DNA: DNA extracted from fossils.

Autogamy: This is defined as a mode of sexual reproduction by fertilization between gametes from the same individual. An individual is said to be allogamous if cross-fertilization is preferential.

Clade: See monophyletic group.

Coalescent analysis: A retrospective approach that aims to reconstruct the genealogical history of a sample of genes, taking into account the demographic history of the population and a mutation model to trace these genes back to their most recent common ancestor.

Diploid (2n): This characterizes a genome made up of pairs of homologous chromosomes (which pair up during meiosis), and, by extension, the corresponding organism.

Domestication syndrome: This includes modification of a set of characteristics during the domestication of plants, such as absence of ginning, increase in the size of harvested organs (seeds or fruits), and clustered flowering and loss of daylength sensitivity.

Exon: This is a segment of a primary RNA molecule conserved during maturation and joined to another exon.

F1 and F2 crosses: In animals, as in yeast, an F1 cross is a cross between two breeds, or two lineages, or two differentiated strains, each individual resulting from this cross receiving, at a given locus, one allele from each parent; an F2 cross is the result of fertilization between two F1 strains. In plants, F1 hybrids result from the crossing of two pure lines, and F2s from the self-fertilization of F1 plants.

F1 hybrids: In plants, the product of crossing two homozygous lines. In many species, such as corn, sunflower, beet, rapeseed and tomato, most varieties cultivated today are F1 hybrids.

Feral: This is a characteristic of a domestic population that has returned to a wild state.

Feralization: This is defined as the return of a domesticated population to a wild state.

Gene flow: The genetic exchanges by hybridization between individuals of similar species or by cross-breeding between individuals of different populations within the same species.

Genome editing: This includes engineering modification of the sequence of a given site of a genome by addition, deletion or substitution of nucleotides, performed or initiated by enzyme complexes targeting this site.

Genomic selection: This is a selection method that takes into account the information provided by molecular markers distributed throughout the genome, making it possible to predict the value of certain traits without having to characterize all individuals for that trait.

Genotyping chip: This is an inert support (such as glass) on which single-stranded DNA fragments are fixed in an orderly fashion, forming a panel of SNPs that can be used to reveal an individual's genotype through DNA hybridization.

GWAS (genome-wide association study) analysis: This is defined as genome-wide study of correlations between variations in specific regions or positions of the genome and variations in certain phenotypic traits measured in a population.

Haploid (n): This refers to a genome composed of a single set of chromosomes and, by extension, to the organism that has them.

Haplotype: This is a set of alleles at distinct loci on the same chromosome, transmitted together. It also refers to the mitochondrial DNA type of an individual in a population.

Heritability: In the broad sense, the proportion of variance measured for a trait that is of genetic origin, and in the strict sense, the proportion of this variance due to additive genetic effects, which is the one estimated by usual methods. By construction, it varies between 0 and 1.

Holocene: This is a geological period extending from 12,000 years ago to the present day.

Horizontal gene transfer: The transfer of genetic material between two organisms outside of sexual reproduction. It mainly involves microorganisms and plants.

Introgression: Integration of genetic material from one species into another, by hybridization and backcrossing by the host species, or possibly by horizontal gene transfer.

Iso-enzymatic markers: Genetic markers corresponding to different forms of the same enzyme, revealed by electrophoresis (differential migration under the action of an electrical field). Their number is limited to a few dozen.

Locus: Position of a gene or DNA sequence on its chromosome.

Meiosis: A succession of two cell divisions during the formation of gametes, resulting in a reduction in the number of chromosomes to half that of the initial cell. During meiosis, exchanges (recombination) take place between homologous chromosomes.

Microbiota: The system in which all of the microorganisms live in a given ecosystem.

MicroRNA: This is noncoding RNA of 21–25 nucleotides that controls gene expression by pairing with target RNAs, with which it is partially complementary.

Microsatellites: DNA sequence containing a variable number of repetitions of a motif of 2, 3 or 4 base pairs.

Mitochondrial DNA: DNA present in the mitochondria, which differs from the nuclear genome contained in the cell nucleus. Mitochondrial DNA is classically used in animal population genetics and evolutionary biology.

MLST (multi locus sequence typing): This is defined as typing of microorganism strains by sequencing a set of DNA fragments, amplified by PCR, from as many essential genes.

Molecular markers: These are DNA sequences, whose variation is used to identify a particular location (locus) on a given chromosome. These markers make it possible to compare genomes between species, varieties, populations and individuals, and to establish relationships between individuals.

Monophyletic group or clade: These are zoological or botanical groups considered to be derived from a single primordial species, from a common ancestor.

Orthologs: This refers to genes present in two different species that have similarities in sequence, because they derive from a common ancestor.

Paleogenetics: The study of the DNA of the ancestors of today's species.

Paleogenomics: The study of the genomes of the ancestors of today's species.

Pan-genome: It is a set of genomic sequences, without distinguishing allele diversity, present in a given species or group of individuals. The pan-genome corresponds to two compartments: the core genome, which groups together sequences present in all individuals; and the dispensable genome, which corresponds to sequences present in one or more individuals, but not in all (anything not present in all individuals (that which is not "core") is dispensable).

PCR (polymerase chain reaction) amplification: This is an in vitro exponential amplification of a defined DNA sequence, involving successive cycles of specific oligonucleotide pairing and polymerase elongation. It enables a very large number of copies of a DNA sequence of interest to be obtained.

Phylogenetics: This is used as an adjective to describe phylogeny, and as a noun to describe the branch of genetics that deals with genetic changes in species during evolution.

Phylogeny: This defines family relationships between present-day species or populations, classically represented in the form of a tree (known as a "phylogenetic tree") or bush in which the groups (represented by the leaves) are on branches that are all the more close that their last common ancestor is recent; a phylogenomic tree is used when whole-genome data are used to reconstruct family relationships.

Polymorphism (of DNA): This is a variation detected in the genomic sequence.

Polyphyletic: A group of individuals or strains that do not derive from a single common ancestor, but from several. See monophyletic.

Polyploid: This describes a nuclear genome (and the corresponding organism) comprising more than two sets of chromosomes.

Promoter: This is defined as a region of DNA near a gene that regulates transcription of that gene.

QTL (quantitative trait locus): This is defined as region of the genome involved in the variation of a quantitative trait, such as a performance level (disease resistance, growth rate, milk yield, etc.).

Selection signature: See selective sweep.

Selective sweep/selection signature: When natural selection favors certain alleles, their frequency increases in the population. Selective sweep is the process by which alleles present in neighboring regions, whether they are advantageous or not, are also swept along by this selection, with their frequency also increasing. This phenomenon and the size of the regions involved depend on the recombination rate between chromosomes of the same pair.

SNP (single-nucleotide polymorphism): This is defined as variation in a nucleotide base pair at a specific locus in a genome, creating a polymorphism between individuals in a species.

SNP panel: This is a set of SNP markers regularly distributed in the genome, enabling the genotype of each individual to be determined at each locus in the panel.

Sympatry, sympatric: Coexistence in the same territory of two closely related species that do not hybridize.

Synonymous mutation (or substitution): Nucleotide change(s) in the sequence of a gene causing a codon change which, due to the redundancy of the genetic code, will not induce an amino acid change, unlike a non-synonymous mutation which causes an amino acid change.

Tetraploid (4n): This describes a nuclear genome (and the corresponding organism) comprising four sets of chromosomes: autotetraploid if it consists of homologous chromosomes, and allotetraploid if it consists of the addition of chromosomes from two diploid species.

Transcription: The process of synthesizing a complementary RNA molecule to a template strand of a DNA molecule.

Transcriptome: All of the RNAs (messenger, ribosomal, transfer, etc.) resulting from genome transcription.

Transcriptomics: Systematic analysis of quantitative and qualitative transcriptome expression using various techniques.

Translation: The process of synthesizing a polypeptide chain, during which the genetic message provided by the sequence of a messenger RNA molecule is converted, according to the genetic code, into the amino acid sequence of a protein molecule.

Transposable element: See transposon.

Transposon: This is defined as DNA fragment capable of moving from one locus to another in the genome, with or without duplication.

List of Authors

Bertrand BED'HOM
ISYEB
Muséum national d'Histoire naturelle
Paris
France

Cécile CALLOU
AASPE
Muséum national d'Histoire naturelle
Paris
France

Mathilde CAUSSE
GAFL
INRAE
Avignon
France

Philippe CUBRY
DIADE
IRD
Montpellier
France

Marguerite DOLS-LAFARGUE
OENO
INRAE
Villenave-d'Ornon
France

Laurent FRANTZ
Ludwig Maximilian University of Munich
Germany
and
QMUL
London
United Kingdom

Virginie GALEOTE
SPO
INRAE
Montpellier
France

Hervé GARREAU
GenPhySE
INRAE
Toulouse
France

Alain GHESQUIÈRE
DIADE
IRD
Montpellier
France

Charlotte HER
LECA
Université Grenoble Alpes
France

Thierry LACOMBE
UMR AGAP Institut
CIRAD
INRAE
Institut Agro
Université de Montpellier
and
UMT Geno-Vigne®
IFV-INRAE-Institut Agro
Montpellier
France

Claire LE HENAFF-LE MARREC
OENO
INRAE
Villenave-d'Ornon
France

Jean-Luc LEGRAS
SPO
INRAE
Montpellier
France

Mathias LORIEUX
DIADE
IRD
Montpellier
France

Patrick LUCAS
OENO
INRAE
Villenave-d'Ornon
France

Cécile MARCHAL
Unité expérimentale du Domaine de Vassal
INRAE
Marseillan
France

Thibault NIDELET
SPO
INRAE
Montpellier
France

Morgane OLLIVIER
ECOBIO
Université de Rennes
France

Georges PELLETIER
Académie des sciences
and
Académie d'agriculture de France
Paris
France

François POMPANON
LECA
Université Grenoble Alpes
France

Caroline PONT
GDEC
INRAE
Clermont-Ferrand
France

Xavier ROGNON
GABI
INRAE
AgroParisTech
Université Paris-Saclay
Jouy-en-Josas
France

Jana RUDOLF
OENO
INRAE
Villenave-d'Ornon
France

François SABOT
DIADE
IRD
Montpellier
France

Jérôme SALSE
GDEC
INRAE
Clermont-Ferrand
France

Delphine SICARD
SPO
INRAE
Montpellier
France

Fabrice TELETCHEA
URAFPA
Université de Lorraine
Nancy
France

Patrice THIS
UMR AGAP Institut
CIRAD
INRAE
Institut Agro
Université de Montpellier
and
UMT Geno-Vigne®
IFV-INRAE-Institut Agro
Montpellier
France

Michèle TIXIER-BOICHARD
GABI
INRAE
AgroParisTech
Université Paris-Saclay
Jouy-en-Josas
France

Index

H, I

L, M

N, O

P

Printed and bound by CPI Group (UK) Ltd, Croydon, CR0 4YY

13/09/2026

14961069-0001